Cell Biology, Genetics, and Evolution

Biology Lab Manual

Second Edition

Jennifer Schramm

Chemeketa Press | Salem, Oregon

Cell Biology, Genetics, and Evolution: Biology Lab Manual
© 2024, 2025 by Jennifer Schramm

ISBN-13: 978-1-955499-37-8

All rights reserved. Edition 1 2024. Edition 2 2025.

No part of this book may be reproduced or transmitted in any form or by any means, electronic or mechanical, including photocopying, recording, or by any information storage and retrieval system, without permission in writing from the publisher.

Chemeketa Press
Chemeketa Community College
4000 Lancaster Dr NE
Salem, Oregon 97305
collegepress@chemeketa.edu
chemeketapress.org

Cover design by Ronald Cox
Interior design by Ronald Cox and Abbey Gaterud

A full list of credits appears on pages 158 and constitutes an extension of this copyright page.

References to website URLs were accurate at the time of writing. Neither the author nor Chemeketa Press is responsible for URLs that have changed or expired since the manuscript was prepared.

Printed in the United States of America.

Land Acknowledgment
Chemeketa Press is located on the land of the Kalapuya, who today are represented by the Confederated Tribes of the Grand Ronde and the Confederated Tribes of the Siletz Indians, whose relationship with this land continues to this day. We offer gratitude for the land itself, for those who have stewarded it for generations, and for the opportunity to study, learn, work, and be in community on this land. We acknowledge that our College's history, like many others, is fundamentally tied to the first colonial developments in the Willamette Valley in Oregon. Finally, we respectfully acknowledge and honor past, present, and future Indigenous students of Chemeketa Community College.

Contents

	Lab and Safety Regulations	5
1	Microscopes and Cells	7
2	Diffusion and Osmosis	25
3	DNA Fingerprinting	41
4	DNA Structure, Replication, and Mitosis	55
5	Gene Expression and Mutations	73
6	Meiosis and Sexual Reproduction	95
7	Genetics	113
8	Natural Selection	135
	Acknowledgments	158

Lab and Safety Regulations

Personal Behaviors
1. Eating and drinking are prohibited in the labs.
2. Wash your hands before leaving the lab.
3. Wash the lab counter & related work areas with disinfectant before and after the lab.
4. Stow your personal items in the cubbies or under your station.
5. Keep the lab counters and work areas uncluttered.
6. Clean-up is your responsibility:
    a. Return cleaned items back to lab kits.
    b. Wash slides and return to the original location (unless told otherwise). Cover slips may be thrown away.
    c. Wash glassware and return to original location (unless told otherwise)
    d. Ensure any lab waste is disposed of in the approved container(s).
    e. Ensure sinks and counters are as clean as when you arrived.
7. Clothing must be appropriate for the lab to be performed.
8. Read the lab in advance so you're aware of proper safety protocols.
9. If you have questions about safety, ask before you do.

Safety Protocols
1. Know where the lab safety equipment is located–fire extinguisher, first aid, eye wash, etc.
2. Wear personal safety equipment (goggles, gloves, aprons) as indicated in the lab instructions or by your instructor.
3. Handle all chemicals and biologicals, including stains, below eye level.
4. Do not assume something can just go down the sink or in the trash. Dispose of wastes in appropriate waste containers.
5. If there is a chemical spill:
    a. Get your instructor.
    b. If the spill is on the floor or counter, keep away from it.
    c. If the spill is on you or your clothes, rinse the area immediately with running water.
    d. If you splash chemicals in your eyes, go immediately to the eye wash station, turn on cold water, remove the red caps and lean down so that the water bubbles into your eyes.
6. In case of injury:
    a. Get your instructor.
    b. If the instructor is not available, call campus safety (x5023) on the lab's phone.
7. If the injury appears severe, also call 911
8. In case of fire:
    a. If your clothes are on fire, yell "FIRE" and roll on the floor or use a coat or fire blanket to smother it.
    b. If chemicals or lab equipment is on fire, call out "FIRE" and evacuate the room.
9. If the fire alarm goes off, take your essential items and leave the room. Head to the evacuation area as told by your instructor. Stay with your class and wait for instructions. Do not re-enter the building until an all-clear is given.

Name: _____ Lab Time: _____ Due: _____

1 | Microscopes and Cells Pre-lab

Instructions

- ☐ Read the lab and then follow the instructions to complete this assignment.
- ☐ You may need your textbook and/or other resources to complete this assignment.
- ☐ Pre-labs must be completed before the start of the lab.

1. Review the Biology Lab Safety regulations (p. 5). Indicate your understanding of the regulations by completing the form linked on your course website or by email.

Scan the QR code or visit https://go.chemeketa.edu/microscope to use the virtual microscope to complete this assignment.

2. **Click the "guide" button** and review the sections entitled Overview, Objective Lenses, and Microscope Care.

1. Indicate the formula you would use to determine the total magnification of an image viewed with a microscope.

2. What should you use to clean the lenses of a microscope?

3. **Click the "learn" button** to explore the structure of a microscope. Use the virtual microscope to locate the parts on figure 1.1 and learn the function of the parts listed below.
   » Coarse Focus Knob
   » Fine Focus Knob
   » Diaphragm
   » Mechanical stage adjustment knob
   » Objectives
   » Ocular
   » On/off switch
   » Rheostat (look this up)
   » Stage

Figure 1.1. Microscope Anatomy

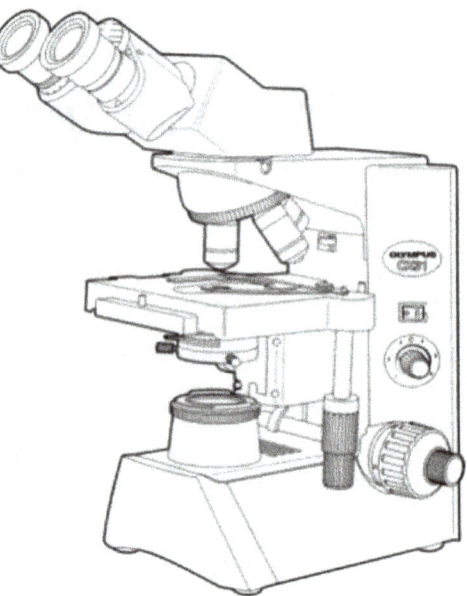

4. Indicate the part of the microscope that you would use in each of the following situations by filling in the "Microscope Part to Adjust" side of table 1.1.

   Table 1.1.

Situation	Microscope Part to Adjust
The specimen is at the edge of your view when you look through the oculars, and you want to center it.	
The image through the oculars is slightly out of focus at 40X magnification.	
The image is in focus, but it is hard to see your specimen due to a lack of contrast.	
The light coming through the oculars is too bright for you to comfortably view the image.	
When preparing to store the microscope, you notice that the high-power objective is very close to touching the stage.	

Figure 1.2. Sketch of Cluster of Cells from *Elodea*
Magnification:

Q1. As you examine *Elodea* with your high power objective, carefully turn the fine focus adjustment knob. What happens?

Adjusting the fine focus knob allows you to look at different layers of cells within the *Elodea* leaf. Although we can place a three-dimensional specimen under the compound microscope, we can only obtain a two-dimensional image of that specimen. In many living specimens, focusing up and down allows us to examine different "slices" of a three-dimensional specimen.

Q2. Estimate the number of layers of cells that form an *Elodea* leaf by focusing up and down on your specimen

Elodea *is an invasive species. Put it in the garbage, not down the sink.*

Exercise 1.2. Viewing Single-Celled Organisms

Life consists of more types of single-celled organisms than multicellular organisms. Single-celled organisms can be quite dynamic and fun to examine! At the same time, they can be small, hard to find, and very fast.

Procedure

1. Create a wet mount using the sample provided by your instructor. Note that since these samples are already in water, you don't need to add additional liquid to the slide.
2. Place the slide on your microscope and focus on the edge of the coverslip. This is the easiest way to make sure you are focusing on the right thing.
3. Use the scanning objective (4x) to find the organisms
4. Slow your organisms down to get a closer look at their structure. Add a very small drop of methylcellulose (Proto-slo or Detain) to the edge of both sides of your coverslip. This substance will slowly diffuse under the coverslip, and the organisms will slow. Too much methylcellulose will destroy all the organisms on your slide.
5. Get a single-celled organism in focus with the 40X objective.
6. Sketch a single-celled organism in the space provided for figure 1.3. Label any cell structures that you can identify.

Figure 1.3. Sketch of a Single-Celled Organism
Magnification:

Q3. Describe several ways that single-celled organisms differ structurally from the plant cells we examined in *Elodea* (Exercise 1.1.C).

Exercise 1.3. Using Stains to Examine Cell Structure

Scientists often stain specimens for observation with the compound light microscope. Stains provide additional contrast, making it easier to see internal structures. In some cases, stains may only attach to specific molecules, allowing scientists to find those molecules within the cell.

A. What Does Iodine Solution Stain?

Iodine solution is commonly used to stain specimens being examined with a compound light microscope because it increases contrast within the cell. In addition, iodine solution reacts with a certain type of carbohydrate to cause a color change. In this activity, you will perform a simple experiment to determine which carbohydrate reacts with iodine.

> *Skills to Remember: How to Add Stain to a Wet Mount*
> 1. Place a drop of the stain on the slide next to the cover slip.
> 2. Twist the corner of a tissue into a point. Place the point at the edge of the other side of the cover slip (opposite the drop of stain).
> 3. Use capillary action to draw the stain under the cover slip.

Q4. What is a carbohydrate, and what role do carbohydrates play in cells?

Q5. Which of the following molecules is/are carbohydrates (check all that apply)
- ❏ Table sugar
- ❏ Olive oil
- ❏ DNA
- ❏ Starch

Procedure
1. Place three watch glasses on top of a piece of white paper.
2. Add a very small scoop (less than a pinch!) of sugar to one watch glass. Simple sugars like this are a short-term form of energy storage in cells.
3. Add a very small scoop (less than a pinch!) of cornstarch to a second watch glass. Cornstarch contains starch, an energy-storage molecule produced by plants.
4. Add several drops of iodine solution to the empty watch glass.

Q6. What color is the pure iodine solution (before adding anything else)?

Q7. This sample is your control. We are observing how iodine reacts with different substances. Why do we examine pure iodine before observing samples of iodine mixed with carbohydrates?

5. Add several drops of iodine solution to the watch glass containing sugar.

Q8. What color does mixing iodine and sugar produce?

6. Add several drops of iodine solution to the watch glass containing cornstarch.

Q9. Q9: What color does mixing iodine and cornstarch produce? _____

Q10. Based on your observations, you should be able to draw a conclusion about how iodine solution can be used in cells. Fill in the blank with your conclusion

Conclusion: Iodine can be used to detect _____ (starch or sugar?) in cells.

7. Create a wet mount using only a few grains of cornstarch and a drop of water.
8. While your slide is still on the microscope, stain your sample with iodine solution using capillary action from before.
9. Reexamine your slide with your microscope.

Q11. What is the iodine actually reacting with on the wet mount you made of cornstarch? The cornstarch or the liquid around the cornstarch?

B. Using Iodine Solution to Stain Plant Cells

Plants store energy both as sugar and starch. In this exercise, you conduct an observational study to determine which plant cells contain starch. In addition, the use of the stain will highlight internal cell structures, which allows you to get a better feel for plant cell structure.

Q12. Before you begin, complete the following statements based on what you think the results will be.

Onion cells contain the energy storage compound _____.

Potato cells contain the energy storage compound _____.

Procedure: Onion Epidermal Cells

1. Follow the instructions provided in the lab to prepare a wet mount of just the epidermis (inner skin) of an onion.
2. Get the specimen in focus on your microscope. Choose the objective that allows you to see individual cells.
3. Stain the specimen with iodine solution (remember not to take it off of the microscope!)
4. Draw several onion epidermal cells in the space provided in figure 1.4.
5. Indicate the total magnification at which you viewed the slide.
6. Label the nucleus, central vacuole, and cell wall of a single cell.

Figure 1.4. Sketch of Onion Epidermal Cells
Magnification:

Q13. How would you account for the fact that some of the nuclei in your onion cells appear to be centrally located in the cells?

Q14. What happens when you add iodine solution to the slide? Describe your observations in detail.

Q15. What type of energy storage compound (sugar or starch) is probably present in onion epidermal cells? How do you know?

Procedure: Potato Cells

1. Follow the instructions provided in the lab to prepare a wet mount of potato cells (the white part).
2. Get the specimen in focus on your microscope. Choose the objective that allows you to see individual cells.
3. Stain the specimen with iodine solution (remember not to take it off of the microscope!)
4. Draw several potato cells in the space provided for figure 1.5.
5. Indicate the total magnification at which you viewed the slide.
6. Label the cell wall and any additional structures you can identify.

Figure 1.5. Sketch of Potato Cells
Magnification:

Q16. Many cells were cut open when you sliced the potato, which released their contents onto the slide. What do you think of the oval objects that are scattered on the slide?

Q17. What happens when you add iodine to the slide? Describe your observations in detail.

Q18. What type of energy storage compound (sugar or starch) is probably present in potato cells? How do you know?

Q19. What do you think is the function of the oval objects that are scattered around the slide? Would you consider these structures to be organelles? Explain.

Q20. Return to the hypotheses you made in Q11. Do your observations of onion and potato cells stained with iodine support your hypotheses? Which cell type stored starch? How do you know?

Exercise 1.4. Observing Animal Cells

While animal cells can also be stained with iodine, we usually use different staining techniques due to the differences in plant and animal cell structure. In this activity, you will use methylene blue, a stain that turns DNA blue and provides an overall increase in contrast in the cell.

Procedure: Human Cheek Cells
1. Follow the instructions provided in the lab to prepare a wet mount of your cheek cells.
2. Get the specimen in focus on your microscope. Choose the objective that allows you to see individual cells.

3. Stain the specimen with methylene blue solution (remember not to take it off of the microscope!)
4. Draw several cheek cells in the space provided for figure 1.6.
5. Indicate the total magnification at which you viewed the slide.
6. Label the nucleus, cytoplasm, and plasma membrane.

Figure 1.6. Sketch of Cheek Cells

Did you observe very small blue dots on the surface of your cheek cells? These are bacteria, prokaryotic cells. Note how much smaller they are than the eukaryotic cells we are focusing on in this lab.

Q21. What part of the cheek cell was most prominently stained by methylene blue? Why do you think this is the case?

Name: _____ Lab Time: _____ Due: _____

Applying What You've Learned in the Microscopes and Cells Lab

7. Compare and contrast the cells you observed during this lab. How are they similar? How are they different? Create a diagram to highlight these differences.

8. Look back through your sketches. At what magnifications were you best able to see larger organelles like the nucleus and chloroplast? Use your observations to explain why it's important to record the magnification used when you make your sketches.

9. Complete table 1.3 comparing the various membrane-bound organelles observed in this lab.

Table 1.3.

Membrane-bound Organelle	Functions	Plant, animal, or all eukaryotic cells?
Nucleus		
Chloroplast		
Mitochondria		
Central Vacuole		

10. In Exercise 1.3, you used a control (watch glass with pure iodine solution). How did you use this control to interpret the results of the experiment performed in this exercise?

11. Describe several roles of stains when using a microscope. Indicate specific exercises in this lab that highlight these roles.

Name: _____ Lab Time: _____ Due: _____

2 | Diffusion and Osmosis Pre-lab

Instructions

- ❏ Read the lab and then follow the instructions to complete this assignment.
- ❏ You may need your textbook and/or other resources to complete this assignment.
- ❏ Pre-labs must be completed before the start of the lab.

1. Differentiate between the processes of diffusion and osmosis.

2. Use the QR code or go to https://openstax.org/books/concepts-biology/pages/2-3-biological-molecules to visit the section of the textbook covering macromolecules and read about carbohydrates. The glucose we are using in Exercise 2.3 in today's lab is a monomer. The starch we are using is a polymer.

   Which of these molecules (starch or glucose) do you think is bigger?
   _____

   Which of these molecules (starch or glucose) is more likely to travel through a selectively permeable membrane? _____

3. Use the instructions for Exercise 2.3 in this lab to sketch the four experimental set-ups your group will create in class. The first one is done for you.

   ☐ Label the contents of each mock cell in figure 2.1.
   ☐ Label the solution that will go on the outside of each mock cell in figure 2.1.

   Figure 2.1. Set up of Osmosis Experiment (Exercise 2.4). Tubes 3 and 4 require heat and contain solutions that must be placed in a special disposal container. *Created with Biorender.*

   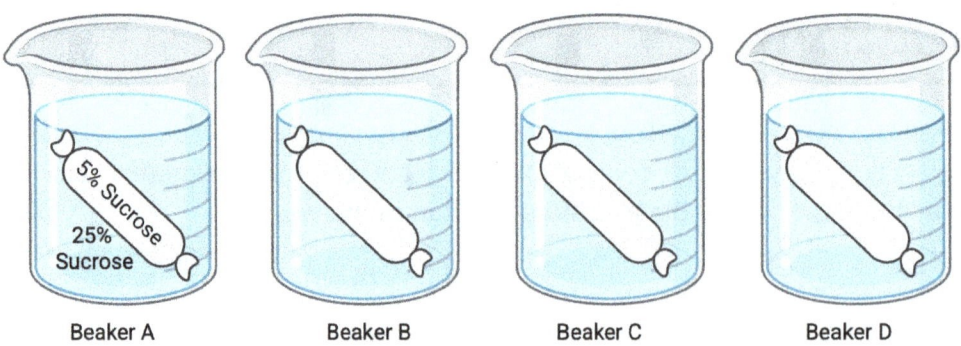

   Beaker A   Beaker B   Beaker C   Beaker D

4. When a hypertonic solution is separated from a hypotonic solution by a selectively permeable membrane, in which direction does water move?

   ☐ Water moves from the _____ solution to the _____ solution.

5. Which chemical used in this lab cannot be disposed of down the sink? Why do you think this is?

Name: _____ Lab Time: _____ Due: _____

2 Diffusion and Osmosis

Understanding how molecules move in and out of cells is key to understanding how cells work.

By the end of this lab, you should be able to:

- ❑ Differentiate between diffusion and osmosis.
- ❑ Explain how size relates to the ability of materials to cross membranes.
- ❑ Describe the relationship between cell shape, size, and function.

Exercise 2.1. Brownian Motion

All molecules are subject to naturally occurring movement. Under the microscope, this movement will appear like a gentle vibration.

A. Procedure

1. Create a wet mount with carmine dye (see Lab 1 for the procedure). Shake the dye first. No additional liquid is required.
2. Place the slide on the microscope and let it sit for a couple of minutes to settle and for the cover slip to level out.
3. Observe the smallest particles on high power.
    a. If you see all the particles on your slide moving across your field of view in the same direction, you should wait a bit longer. The movement you are seeing is the flow of water across the slide. Ignore this movement. When the slide has settled, you will be able to observe the movement of individual particles.

Q1. What are the particles doing?

Q2. Look at a single particle. Does it always move in a single direction? Describe the path it seems to follow.

This movement was first observed by Robert Brown, who concluded that the particles must be moving because water molecules were colliding with them (despite the fact that you cannot see a water molecule).

Exercise 2.2. Diffusion

Diffusion is a process in which molecules and particles move by random collisions and eventually become randomly distributed throughout a solution. The result of this is the apparent movement of those particles from high concentrations toward low concentrations of the substance that is diffusing. We say that the particles are diffusing along a concentration gradient from regions of high concentration to regions of low concentration.

Q3. Do all molecules pass readily through a membrane? What determines which particles will pass through a membrane and which won't?

Procedure
1. **Fill a 100 mL beaker** with water to within 2 cm of the top. Add enough IKI (iodine) solution to turn the beaker contents a medium golden brown. Put about 1 cm of this solution in a test tube for use in the controls section (Tube 1 from table 1).
2. **Obtain a 15 cm piece of soaked dialysis tubing**. Tie one end of the tube into a knot (the knot should be as close to the end of the tubing as possible). Roll the tubing between your fingers to open it up to a bag.
3. Use a dropper to add **one squirt of starch solution** into the bag.
4. Use a dropper to add **one squirt of glucose solution** into the same bag..
5. **Rinse** off the outside of the bag to remove any spills.

6. **Hang the dialysis tubing** over the side of the beaker with the bag hanging in the iodine water. Hang the other end outside of the beaker and use a rubber band to hold it in place.
7. **Set your beaker aside** for about 30 minutes.
8. While you wait, prepare the control test tubes.

Controls

Experiments are both controlled and contain controls. Generally, only one condition is changed at a time (that's the "controlled" part). A control is a treatment in which the condition being tested by the experiment is held constant. Controls act as a baseline allowing scientists to compare the results of experimental treatments to their standard to determine if their treatment actually causes a change. There are many types of controls, including:

- ❑ **Positive control:** shows what a positive result looks like (a definite change).
- ❑ **Negative control:** shows what a negative result looks like (no change).

Prepare the controls for today's experiment

1. Fill a 400 ml beaker with 2–3 cm of water.
2. Place the beaker on a hot plate and bring the water to a boil.
3. Prepare the four test tubes described in figure 2.2 and table 2.1 by adding 20 drops of the required solutions to each tube.
4. Place test tubes 3 and 4 into the boiling water for 5 minutes.
5. Complete table 2.1 before moving on.

> **Benedict's reagent** is a chemical commonly used in medical labs to indicate the presence of glucose in bodily fluids like urine. If glucose is present, you will see a color change.

Figure 2.2. Controls for the Diffusion Experiment. *Created with Biorender.*

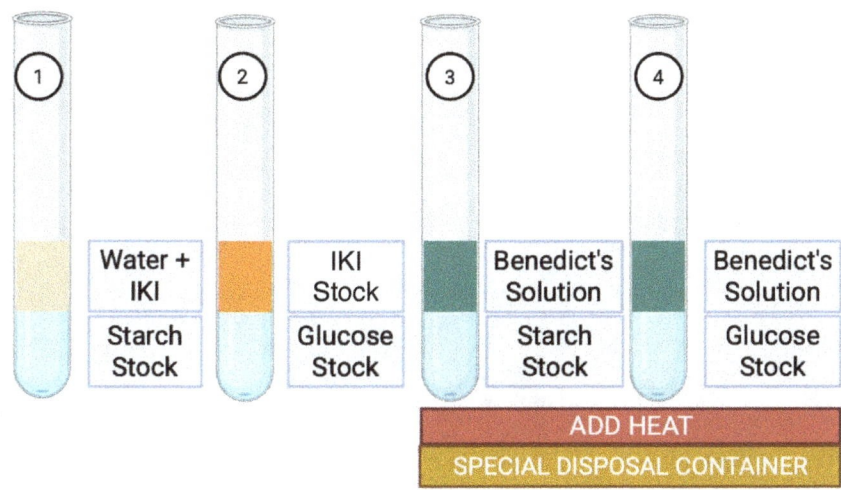

Diffusion and Osmosis 29

Table 2.1. Controls for Diffusion Experiment

Tube #	Add 20 drops of...	And 20 drops of...	Result of Reaction
1	Starch stock solution	IKI + water (Beaker Solution)	
2	Glucose stock solution	IKI stock solution	
3	Glucose stock solution	Benedict's Reagent & Heat	
4	Starch stock solution	Benedict's Reagent & Heat	

Note: Benedict's Reagent cannot go down the drain.

Q4. We are testing whether starch and glucose got out of the bag and whether iodine got into the bag. In each question below, describe any observations that helped you answer the question.

Can you tell if glucose diffused out of the bag?

Can you tell if starch has moved out of the bag?

Can you tell if iodine has moved into the bag?

Q5. Explain how the control test tubes helped you understand these results.

Q6. Write out the step-by-step procedure your group will use to determine if glucose moved out of the bag. *Hint: look at the supplies provided in your kit for this lab!*

Get your instructor's initials before running this experiment. _____

Q7. Has glucose moved out of the bag? Describe any observations that helped you answer this question.

Q8. Draw a diagram in figure 2.3 showing the progress of this experiment. On the left, draw the beaker and dialysis tubing at the start. Label the location of each solution used. On the right, draw the beaker and dialysis tubing at the end. Label the location of each solution used.

Figure 2.3. Set Up of the Diffusion Experiment. *Created with Biorender.*

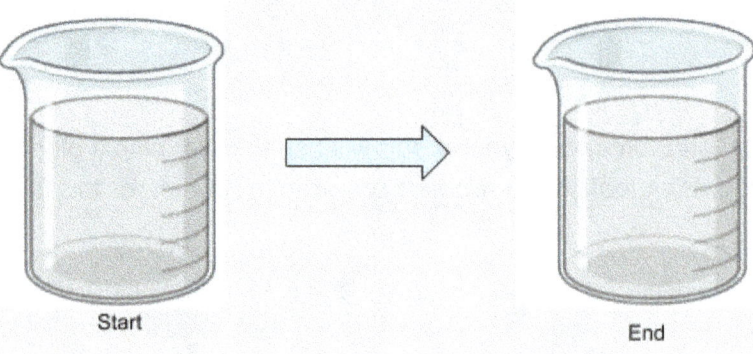

Start   End

Q9. In relative terms, a molecule of iodine is small, a molecule of glucose is medium-sized, and a molecule of starch is very big. What can we infer about the structure of the dialysis membrane we used in this experiment based on the results of our experiment?

6. Clean up! **Do not forget to empty any tube containing Benedict's solution into the provided waste container.**

Exercise 2.3. Osmosis

Water molecules move along concentration gradients just like solute molecules dissolved in the water; however, they move in opposite directions. Water moves from regions of low solute ("high water") concentration to regions of high solute ("low water") concentration. The movement of water across membranes is known as **osmosis**.

*Procedure**

1. Label four 150-ml beakers with the letters A–D.
2. Add 80 ml of the appropriate solution to each beaker (see table 2.2).

   Table 2.2. Set up for Osmosis Experiment

Beaker Letter	Beaker Solution	Mock Cell Solution
A	25% Sucrose	5% Sucrose
B	dH2O	dH2O
C	dH2O	10% Sucrose
D	dH2O	25% Sucrose

3. Soak 4 pieces of dialysis tubing in distilled water for a minute or more.
4. Tie the tubing in a knot about 2 cm from one end. Open the other end of the tubing by rubbing it together between your fingers.
5. Fill the bags about half full with the solutions indicated in table 2.2.
6. After filling the bag, place the appropriate label inside.
7. Force as much air out of the bag as possible by placing it between two fingers and moving your hand away from the solution in the tubing. Tie the open end of the tube in a knot. The bag should be limp after it is tied.
8. Squeeze your bag gently to check for leaks. If you find a leak, re-tie your bag.
9. Rinse your bag with tap water and blot it on a paper towel to remove excess water.
10. Place a weight boat on a digital scale and press the zero or "Tare" button. This will set the weight of the scale (and the weight boat) to zero. Make sure the scale is set to measure grams (g).
11. Place your "cell" in the weigh boat and record its weight to the nearest 0.1 g in table 2.3 in the column labeled "Weight at beginning of the experiment."
12. Repeat until all 4 mock cells are prepared and weighed.
13. When all the bags for your group are ready, place each bag into the appropriate beaker.
14. Wait for 1 hour before proceeding with the experiment.
15. While you wait, make a prediction about how the weight of each bag will change over the course of the experiment in the column labeled "Predicted weight change" in table 3. Use +, -, or 0. You do not need to predict the amount of weight a bag might gain or lose.
16. After 1 hour, remove the bags from their solution and blot excess water away from the bag with a paper towel.
17. Weigh the bag to the nearest 0.1 g and record your results in table 2.3 in the column labeled "Weight at end of experiment".
18. Indicate weight gain for each bag with a "+," weight loss with a "-," and no change in weight with a "0" in the column labeled "Final Change in Weight."

---

* Procedure modified from Glick et al., *The Process of Science: Seven Studies of Life*

Table 2.3. Weight Change Over Time During the Osmosis Experiment

	Contents of Bag	Contents of Beaker	Predicted weight change (+, –, 0)	Weight at beginning of experiment (grams)	Weight at end of experiment (grams)	Final change in weight (+, –, 0)
A	5% Sucrose	25% Sucrose				
B	Distilled Water	Distilled Water				
C	10% Sucrose	Distilled Water				
D	25% Sucrose	Distilled Water				

19. Place the 25% sucrose solution from Beaker A into the "Recycled 25% Sucrose" container. All other solutions can go down the drain. The dialysis tubing can be placed in the garbage.

Q10. Do any of the samples in this experiment represent negative controls? Explain your reasoning.

Q11. What caused the change in weight of your mock cells in this experiment? Is the change the result of the movement of sugars or the movement of water? Explain your answer.

Q12. In table 2.4, explain the weight change in each bag based on the principles of osmosis. If the weight of a bag did not change, explain why. In your explanation, use the terms hypertonic, hypotonic, and isotonic. Drawing pictures here first may help you better explain why the weight changed in your mock cells.

Table 2.4.

Bag	Explanation of Results
A	
B	
C	
D	

Exercise 2.4. Osmosis in *Elodea* (Instructor Demo)

The plant cell wall and the central vacuole play important roles in regulating osmosis in plants. Water that enters a plant cell through osmosis is stored in the central vacuole. As the central vacuole swells with water, it pushes against the cytoplasm and the cell wall, creating the turgor pressure that keeps plants "upright."

Your instructor has set up two microscopes with samples of *Elodea*:

- ❏ *Elodea* in pure water.
- ❏ *Elodea* in a highly concentrated salt solution.

Use your observations to determine which is which.

Q13. Which microscope (your instructor will have them labeled microscope A and B) contains *Elodea* bathed in a high salt solution? How do you know this? Explain your reasoning using the terminology appropriate to discussions of osmosis.

Q14. If you were to immerse an entire *Elodea* plant in salt water instead of just a few cells, how would the plant change?

Name: _____ Lab Time: _____ Due: _____

2 | Applying What You've Learned in the Diffusion & Osmosis Lab

1. Review the results of Exercise 2.1. Why do the dye particles you observed move? How does this movement relate to the processes of osmosis & diffusion?

2. Think back to Exercise 1.3 with starch, sugar, and iodine in Lab 1. We put iodine on one watch glass, iodine and sugar on a second watch glass, and iodine and starch on a third watch glass. Then, we observed the effect of iodine on plant cells.

    ❑ Which of the watch glass samples was the positive control? _____
    ❑ Which of the watch glass samples was the negative control? _____

3. When evaluating the validity of an experiment, scientists always consider the control. Results of experiments that have poorly designed controls or none at all are often questioned or rejected. Why are controls so important?

Diffusion and Osmosis 37

4. Complete table 2.5 to indicate the role of each test tube in the experiment performed in Exercise 2.3.

Table 2.5.

Tube #	Contents	Type of Control (Positive or Negative?)	Summary of Results
1	Starch + Iodine		
2	Glucose + Iodine		
3	Glucose + Benedict's Reagent & Heat		
4	Starch + Benedict's Reagent & Heat		

5. Sketch a diagram of the results of Exercise 2.3, showing which direction each solute (glucose, iodine, starch) moved over the course of the experiment.

6. In figure 2.4, use your explanations of the results of Exercise 2.4 (Q13) to diagram the outcome of the experiment for each beaker.

Figure 2.4. *Made with Biorender.*

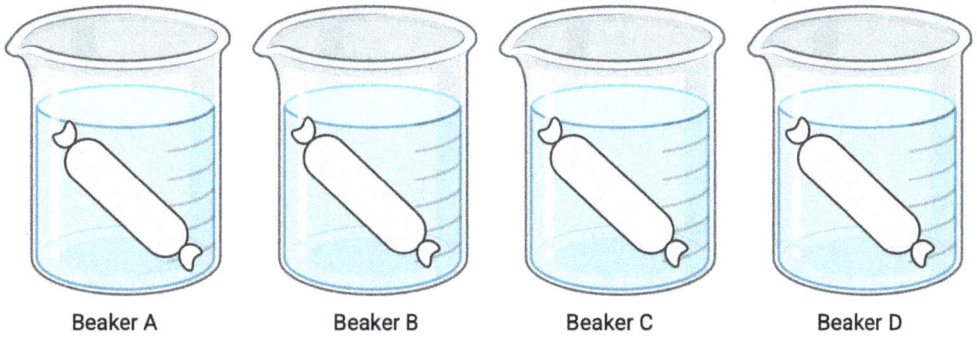

Beaker A   Beaker B   Beaker C   Beaker D

7. In Exercise 2.5, you observed the impact of osmosis on plant cells. Animal cells do not have a cell wall. How do you think the shape of an animal cell would be affected by osmosis?

Name: _____ Lab Time: _____ Due: _____

3 | DNA Fingerprinting Pre-lab

Instructions

- ☐ Read the lab and then follow the instructions to complete this assignment.
- ☐ You may need your textbook and/or other resources to complete this assignment.
- ☐ Pre-labs must be completed before the start of the lab.

Watch the Amoeba Sisters' "Gel Electrophoresis" video by scanning the QR code or visiting https://go.chemeketa.edu/gelelectrophoresis and then answer the following questions:

1. In what organisms are restriction enzymes normally found, and what do they do for that organism?

2. How do scientists use restriction enzymes?

3. What is the charge of a molecule of DNA?

4. During agarose gel electrophoresis of DNA:

    ❑ What causes the DNA to move through the gel?

    ❑ Based on what characteristic of the DNA are pieces of DNA separated?

5. When looking at an agarose gel of DNA, where would the biggest pieces of DNA be located? The smallest?

6. Most of the solutions we use in this week's lab will be dispensed in microliters. How big (or small) is a microliter? How would you describe the volume of this much solution to someone?

Name: _____ Lab Time: _____ Due: _____

3 DNA Fingerprinting

All individuals (except identical twins) have unique DNA sequences. As we will soon learn in lecture, this uniqueness is a result of meiosis and sex. Using unique DNA sequences to identify paternity, criminals, or unidentified bodies has become commonplace in our society. In this lab, we take a closer look at how this process is performed.

By the end of this lab, you should be able to:

- ❑ Describe how and where the DNA of different individuals differs.
- ❑ Explain the origin and function of restriction enzymes.
- ❑ Describe the process of agarose gel electrophoresis.
- ❑ Explain how DNA fingerprinting can be used to differentiate between individuals.

Exercise 3.1. DNA Fingerprinting

A. Practice Loading a Gel

Your instructor will demonstrate the use of a micropipette and the process of loading a gel before distributing a tube of practice dye (PD) and a reusable silicone practice gel.

Procedure
1. Fill the dish containing your practice gel with tap water until the gel is fully covered by the water. The water should be deep enough that its surface is mirror-smooth.
2. Set a micropipette to 10 microliters and add a tip. Everyone in your group will use the same tip.
3. Use the micropipette to suck up 10 microliters of practice dye.
4. Slowly drop the dye into the well as described by your instructor.
5. Everyone in your group should load at least one well on the practice gel.

After everyone has practiced:
- ❑ Return the PD tube to your instructor (assuming there is remaining practice dye).
- ❑ Rinse off the silicone practice gel and its container and return them to your student kit.

B. Gel Loading, Running, and Viewing

Procedure

1. Place your agarose gel in the electrophoresis apparatus. Make sure the wells of the gel are near the black (negative) electrode.
2. Place the electrophoresis apparatus near the power source that you will be sharing with the other team at your lab bench. Be sure that the cables on your apparatus reach the power source, as you cannot move the gel once the samples have been loaded.
3. Add enough 1X TAE to cover the surface of your agarose gel. The solution should fill the wells on both sides of the gel and cover the gel. The surface of the solution should be smooth. Check with your instructor before continuing.
4. Place your samples in the micro-centrifuge directly across from each other. Use the pulse button to spin them briefly (10 seconds).
5. Set your micropipette to 18 microliters.
6. Using a different tip for each sample, load your DNA samples into the gel.
    a. Load the CS sample into the first lane,
    b. Skip a lane
    c. Load your suspects in order in the remaining lanes.
7. Carefully slide the cover onto the electrophoresis apparatus and plug your gel into the power source.
8. When the other team using your power source has attached their gel, turn on the power to high. If everything is set up correctly, you should see tiny bubbles streaming up the short sides of the apparatus.
9. Allow the gel to run for 15–30 minutes. You will know it is done when the low band of stain (light blue) is 3–5 cm away from the wells.
10. After the gel has finished running, remove it from the electrophoresis apparatus and carefully carry it to one of the viewing stations around the room.
11. Open the lid of the Blue View Transilluminator and place your gel on the blue surface. Close the lid so that the transparent orange plastic is between you and your gel.
12. Turn on the transilluminator by turning the knob on the right-hand side. Your bands will become visible.
13. Draw or take a picture of your gel to determine the banding pattern for each lane.

Exercise 3.2. Understanding Restriction Enzymes and Electrophoresis

DNA fragments are separated by size during electrophoresis. One way of measuring DNA size is to determine how many base pairs it contains. Figure 3.1 diagrams a short fragment of DNA. Remember that DNA has two sugar-phosphate backbones where nucleotides are attached to each other by covalent bonds. The two resulting strands attach to each other with hydrogen bonds between their bases. The two bases together are called a base pair (bp).

Figure 3.1. DNA Fragment with Numbered Base Pairs (bp). *Created with Biorender.*

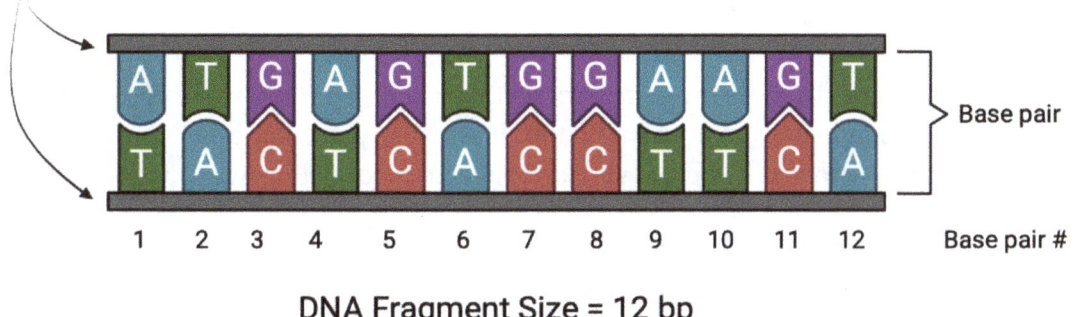

When discussing "pieces" of DNA, scientists will name them by the number of base pairs they contain. In figure 3.1, the base pairs are labeled sequentially from left to right (much like mile markers on a highway). The piece of DNA we see here has a size of 12 bp.

Q1. Looking at figure 3.1, imagine that a restriction enzyme cuts that fragment between base pair 3 and 4. How big would the TWO resulting DNA fragments be?

Figure 3.2 contains a map of a fictional DNA fragment where the **restriction sites** for two **restriction enzymes**, BamHI and HindIII are labeled. Keep in mind that a restriction site represents the sequence of DNA recognized by the enzyme, but the enzyme only cuts the DNA once on each of the two sugar-phosphate backbones. So, the enzyme BamH1 recognizes a region at both 1246 and 7592 bp and cuts specifically at those locations—no other locations on the map are affected.

Figure 3.2. Map of a DNA Fragment with Restriction Sites

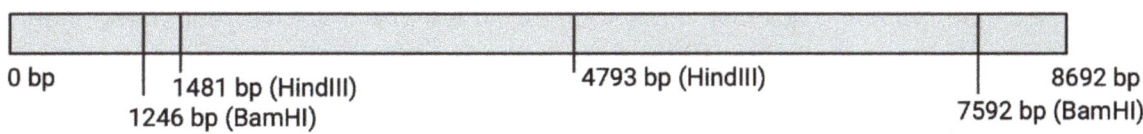

BamHI and HindIII are restriction enzymes. This DNA fragment is 8,692 base pairs long. The labels represent the number of the specific locations (in base pairs) where the sugar-phosphate backbone is cut by that enzyme (the **restriction site**).

In this exercise, we will predict what an agarose gel would look like if this fictional fragment were treated with each specific enzyme on the map, individually, and by a combination of all both restriction enzymes listed at the same time.

Work through the questions below to determine the size of the fragments produced by each restriction enzyme (and both at the same time). Make sure you show all of your work so that your instructor can help you if you get stuck or have an incorrect answer.

Q2. How many fragments would you expect to see on a gel if you treated this DNA fragment with BamHI? What size is each fragment? Show your work.

Q3. How many fragments would you expect to see on a gel if you treated this DNA fragment with HindIII? What size is each fragment? Show your work.

Q4. How many fragments of DNA would you expect to see on a gel if you digested this DNA fragment with both BamHI and HindIII? What is the size of each fragment? Show your work.

Now that we have determined the size of the DNA fragments produced by each restriction enzyme digestion, we can predict what the agarose gel will look like. Use your answers on the previous page to "build" your gel. Remember:

- ❏ The DNA loaded into a particular lane travels vertically from the wells to the bottom of the gel.
- ❏ Each different-sized DNA fragment produced by a restriction enzyme digest appears as a unique band in that lane of the gel. See lane 1 in figure 3.3.

- ❑ DNA fragments placed in an electrical field separate by size with large fragments staying close to the well and smaller fragments traveling down the gel.
- ❑ Two fragments of the same size in different lanes will travel the same distance on the gel.

Most scientists use a **DNA ladder** to help them interpret the results of agarose gel electrophoresis. The DNA ladder includes pieces of DNA of known sizes (specific number of base pairs). Since DNA pieces of the same size travel the same distance in the gel, a DNA ladder can be used to estimate the size of unknown DNA fragments in other lanes. In this case, the DNA ladder in figure 3.3 includes the following sizes of DNA: 9,000 bp, 6,000 bp, 3,000 bp, 1500 bp, and 300 bp.

Figure 3.3. Predicted Results of Agarose Gel Electrophoresis of Fictional DNA Fragment Digested with Different Restriction Enzymes. *Created with Biorender.*

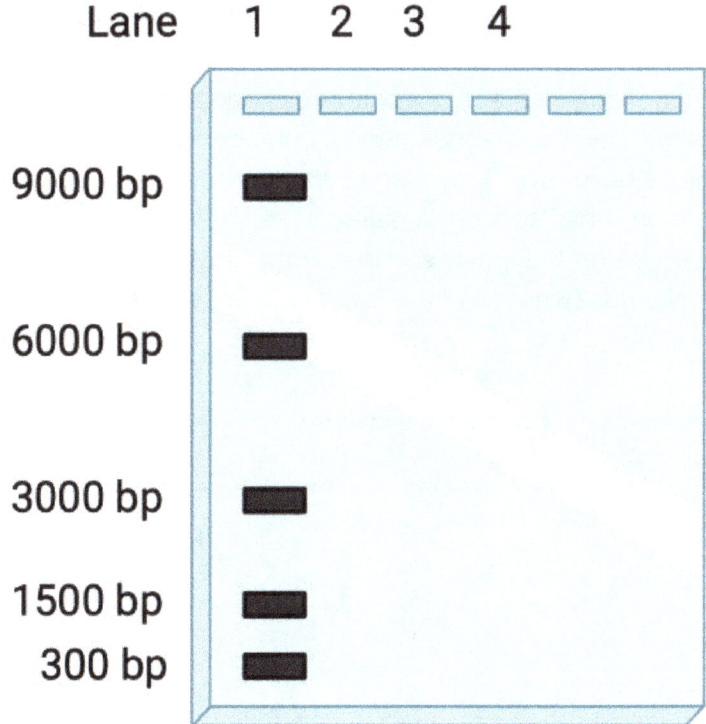

Load the gel as follows:

- ❑ Lane 1: DNA Ladder
- ❑ Lane 2: DNA + BamHI
- ❑ Lane 4: DNA + HindIII
- ❑ Lane 5: DNA + BamHI & HindIII

1. Complete the gel based on the calculations you made in Q2-Q4.
2. Label each fragment with its size (write the size on or next to the bar you draw in the gel).
3. Also, on the gel below, how you would orient the positive and negative poles of the electric field in order to separate the DNA fragments?

Q5. In the DNA fingerprinting lab, you will use electrophoresis of DNA to figure out "whodunit." Explain how you will use your agarose gel to make this determination.

Now that you understand how DNA fingerprinting works, answer the following questions.

Tracy and John had a son, Tyler, in 2009. Two years later, the couple decided to separate. Soon after their separation, Tracy discovered she was pregnant again. John denies that he is the father of Tracy's second child because he knows Tracy started dating Steve soon after the separation. To determine who should pay child support, a paternity test is performed. Blood samples are collected from each member of the family and also from Steve. The DNA is digested with a combination of restriction enzymes, and the fragments are separated by gel electrophoresis. The gel is depicted in figure 3.4.

Figure 3.4. Using Restriction Enzyme Digests and Agarose Gel Electrophoresis to Determine Paternity. *Created with Biorender.*

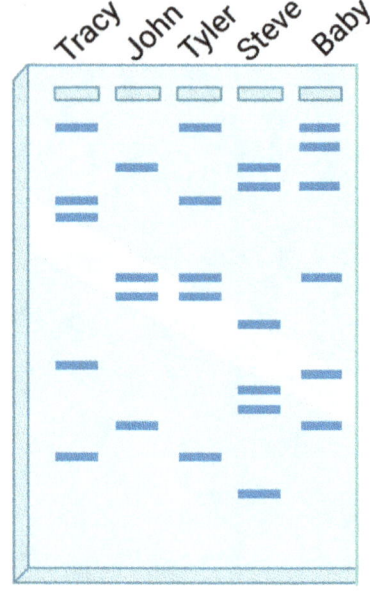

Q6. Who is the father of Tracy's second baby? Explain your answer.

Q7. If the DNA fingerprint of Tracy's first son, Tyler, was not provided on the gel, would you be able to determine the father of Tracy's baby? Why or why not?

Q8. What exactly does a DNA fingerprint represent?

Exercise 3.3. Isolating DNA from Strawberries

Although we cannot see individual molecules of DNA at the molecular level, we can isolate large quantities of DNA by breaking open large numbers of cells. The goal of this laboratory exercise is for you to see, touch, and feel DNA. Scientists in biotechnology and research laboratories perform essentially the same technique you are performing today to study DNA in action.

How this procedure works:

- **Mashing:** this process breaks down the cell walls of the fruit.
- **Extraction solution:** this solution contains detergent, salt, and water.
    » The detergent breaks down membranes.
    » DNA carries a negative charge and is therefore very soluble in water, so when the cells are broken open, the DNA goes into solution.
    » The salt (NaCl) breaks down into ions in the water. Positively charged sodium ions (Na+) are attracted to and bind the negatively charged DNA molecules. This makes the DNA molecules neutral, so they clump together.
- **Ethanol:** DNA is not soluble in ethanol. When ethanol is added to the solution, the DNA can no longer stay in the solution and forms a precipitate.

Q9. What two membranes found in the eukaryotic cell need to be broken to release DNA?

Procedure

1. Place a strawberry in a sealable plastic bag.
2. Squeeze out any extra air and seal the bag.
3. Place the bag on the counter and use the heel of your hand to thoroughly smash the strawberry.
4. Add a volume of DNA extraction solution equal to the volume of strawberry already in the bag.
5. Reseal the bag and place it into a hot water bath for 15 minutes.
6. Transfer the bag to your ice water bath for 5 minutes.
7. Set up the filtration system.
    a. Place a funnel over a clean 500 ml beaker.
    b. Insert a coffee filter into the funnel.

8. Pour your cooled extraction solution (liquid only) into your filtration system and allow it to drip through into the beaker. It may take several minutes for all the liquid to pass through the filter.
9. Dispose of the bag and fruit debris in the garbage can.
10. When filtration is complete, dispose of the coffee filter in the garbage can.
11. Swirl the solution in your beaker. Immediately after swirling, transfer the filtrate into a test tube to a depth of 2-3 cm.
12. Carefully pour a layer of about the same amount of cold ethanol on top of your filtrate in the test tube. The ethanol must be very cold, so be sure to keep it on ice at all times!
13. Let the solution sit for 2 minutes.

A clear gelatinous precipitate (snot) will form between the alcohol layer and the water layer (extraction solution). The white stuff is a clump of DNA from the cells of the strawberry. Use a plastic loop to drag the DNA up the side of the test tube and into a watch glass. It's safe to touch, so feel free to check it out!

14. All materials in this laboratory can be disposed of in the garbage or down the sink.

Q10. A similar procedure can be used to extract DNA from your own cheek cells. In what ways would human DNA be like the DNA you just extracted from strawberries?

Q11. Your precipitated DNA contains both nucleic acids and proteins. What is the job of the proteins that are tightly associated with DNA?

Q12. The nucleus of every human cell contains approximately 2 meters of DNA. A typical adult human contains 60 trillion cells (60,000,000,000,000). The distance from the Earth to the Moon is 380,000 km. If the DNA of a single human were laid end to end, how many times would the strand go to the moon and back?

Name: _____ Lab Time: _____ Due: _____

3 | Applying What You've Learned in the DNA Fingerprinting Lab

1. Why do scientists tend to focus on the non-coding "junk" DNA when performing DNA fingerprinting?

2. Explain why restriction enzymes produce different-sized DNA fragments from the isolated DNA of different people.

3. Examine figure 3.5, an agarose gel from a Biology 102 student. Based on this gel, which of the suspects most likely committed the crime?

Figure 3.5. Sample DNA Fingerprinting Gel from BI102. CS = crime scene. S1–S5 = suspects 1–5. *Created with Biorender.*

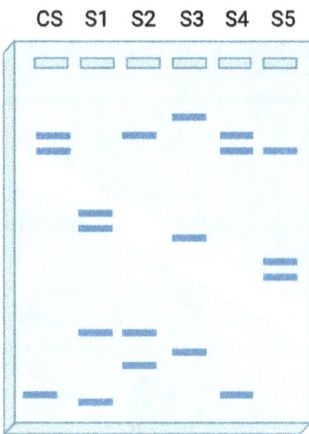

4. Based on what you learned about DNA variability and agarose gel electrophoresis, explain why the suspect in figure 3.5 is the most likely criminal.

5. Examine the agarose gel that your group created during the lab. Based on the patterns of DNA visible in that gel, which suspect committed the crime?

Name: _____ Lab Time: _____ Due: _____

4 | DNA Structure, Replication, and Mitosis Pre-lab

1. The diagrams in figures 4.1 and 4.2 include two different structures. One is a **nucleotide**, and the other is a **piece of DNA**.

    a. Label the diagrams in figures 4.1 and 4.2 with the name of the structure ("nucleotide" or "piece of DNA").
    b. Nucleotides contain three parts. Label those three parts in the diagram.
    c. DNA contains a **sugar-phosphate backbone** and nucleotide **base pairs**. Label those parts in the diagram.

    Figure 4.1. *Created with Biorender.*

    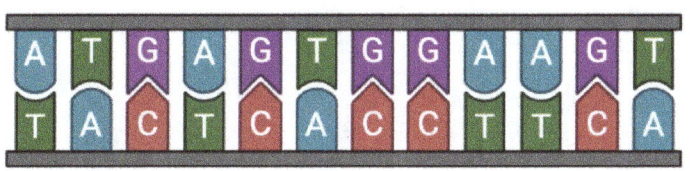

    Name of Structure:

    _____

    Figure 4.2. *Created with Biorender.*

    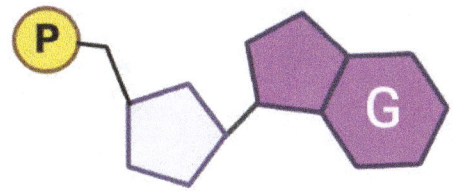

    Name of Structure:

    _____

2. What do the As, Gs, Ts and Cs in figure 4.1 and the G in figure 4.2 represent?

3. Figure 4.3 represents the cell cycle. Interphase has been labeled to provide a point of reference. Complete the diagram by labeling the parts of the cell cycle: **cell division, cytokinesis, G1 phase, G2 phase, S phase**.

Figure 4.3. Diagram of the Cell Cycle. *Created with Biorender.*

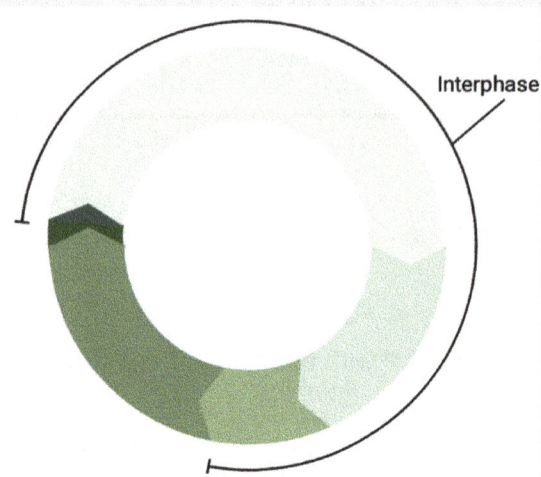

4. What are the two types of cell division that may occur during the cell cycle?

5. Complete table 4.1 comparing the different stages of mitosis. Some boxes have been completed for you.

Table 4.1.

Phase	Nucleus (absent or present?)	Chromosomal Arrangement
Prophase		
Metaphase		
Anaphase		
Telophase		Sister chromatids reach opposite spindle poles and begin to decondense

Name: _____ Lab Time: _____ Due: _____

4 | DNA Structure, Replication, and Mitosis

By the end of this lab, you should be able to:

- ❏ Describe the structure of a DNA molecule.
- ❏ Explain how DNA structure facilitates DNA replication.
- ❏ Describe what happens during each stage of the cell cycle.
- ❏ Use chromosome models or drawings to explain the process of mitosis to another student.
- ❏ Identify the stage of the cell cycle of a dividing cell.

Exercise 4.1. The DNA Puzzle

In this exercise, we will examine DNA structure and the process of DNA replication in detail. We will use a DNA puzzle kit with cardboard pieces to model the structure and replication of DNA. In order to understand the events in DNA replication it is important to work both with the correct pieces and assemble them in the order described. Resist the temptation to skip steps.

A. The DNA Nucleotide

Examine the **deoxyribose** (dark pink) pieces from the puzzle.

- ❏ Deoxyribose is a five-carbon ring sugar.
- ❏ Individual carbon atoms are not shown on the puzzle piece but are understood to be at the corners of the pentagon (see drawing of deoxyribose on the right).
- ❏ The fifth carbon is outside the ring, located counterclockwise from the oxygen atom in the ring.
- ❏ The three knobs on the puzzle piece represent sites where covalent chemical bonds can form.
- ❏ The carbons are numbered in a clockwise direction, as shown in figure 4.4.

Figure 4.4. Chemical Structure of Deoxyribose

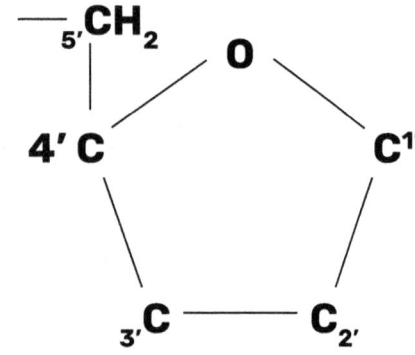

Deoxyribose

Procedure

2. **Build 24 DNA nucleotides**. For each nucleotide:
    a. Attach a yellow phosphate to the 5' carbon on the deoxyribose.
    b. Attach a base (adenine, guanine, cytosine, or thymine) to the 1' carbon.
    c. Set aside **half of each type of nucleotide** for use later.

Q1. What type of organic compound (carbohydrate, lipid, protein, or nucleic acid) is deoxyribose?

Q2. Once a base has been attached to a sugar-phosphate, what do we call the 3-part molecule that results?

Q3. What are the 4 types of DNA nucleotides produced in this puzzle?

B. Base Pairs

Specific bases are said to be **complementary** to each other. The exposed end of the base in a nucleotide has a distinctive shape. The exposed edges of complementary bases will "fit together." The bases in your puzzle also have dotted lines extending off of the piece on the side opposite the sugar attachment. These represent potential hydrogen bonding sites that can pair up with a complementary base on the other side of the molecule.

Q4. Which base in the puzzle "fits" with Adenine? _____

Q5. Which base in the puzzle "fits" with Guanine? _____

Q6. Which base in the puzzle "fits" with Cytosine? _____

Q7. Which base in the puzzle "fits" with Thymine? _____

Q8. Does any of the bases pair with itself?

Procedure
1. Arrange the nucleotides into groups containing one type of nucleotide.

C. Building One Side of a DNA Molecule

One side of a DNA molecule is formed by the covalent bonding of nucleotides to one another through the phosphate of one and the sugar of the next. Covalent bonds between atoms are strong and difficult to break.

Procedure
1. Select an **adenine** nucleotide and a **guanine** nucleotide.
2. Attach the **phosphate of the guanine** nucleotide to the **sugar of the adenine** nucleotide.
3. Attach a **phosphate of a thymine** nucleotide to the **sugar of the guanine** nucleotide.

You should now have a chain of 3 nucleotides. Let's add a few more nucleotides.

4. Attach a **cytosine** to the end of your chain of three nucleotides.

5. Attach a **guanine** to the cytosine
6. Attach a **cytosine** to the guanine.

The exposed phosphate end of the chain is called the **5′ (5-prime)** end because the #5 carbon of the sugar is closest to this end. The unbonded exposed sugar end is the **3′ (3-prime)** end because this is the #3 carbon.

Q9. What is the sequence of the DNA fragment you produced? (provide a sequence of letters)

Q10. Use your model to explain why one side of a DNA molecule is called the sugar-phosphate backbone.

Get your instructor's initials before moving on. _____

D. Building a Complete Molecule of DNA

A molecule of DNA contains two strands of DNA nucleotides held together by the hydrogen bonds between complementary bases. We will now convert your single strand into a double-stranded DNA molecule by pairing the correct nucleotides.

Procedure

1. Starting at the adenine end of your strand of nucleotides, attach the matching nucleotide by joining their bases.
2. Add the base complementary to the next nucleotide in your strand (guanine).
3. Form a strong covalent bond between these new bases by linking their sugar-phosphate backbones.
4. Continue until all the bases in your original strand have partners.

Note that the 3′ end of one side is opposite the 5′ end of the complementary side. The two backbones are equidistant from one another but run in opposite directions.

E. DNA Replication

DNA replication requires that the two nucleotide strands of the molecule be separated from one another, and then each strand is copied separately. The processes of separation and copying are both controlled by enzymes. The enzyme that separates the two strands does so by breaking the weak hydrogen bonds between bases. **DNA polymerase** is the enzyme that attaches to each side of the open DNA and travels along, pairing new nucleotides with the existing strand. Due to its structure, DNA polymerase can only move in one direction on a strand of nucleotides—from the 3' end toward the 5' end of the strand it is reading.

Procedure

1. Arrange the remaining 12 nucleotides (built in part A) into 4 groups based on their type.
2. Slide the two sides of your DNA molecule apart from one another to a distance of 6 inches.
3. Identify the 3' end of one nucleotide strand
    a. Copy one strand of the DNA molecule
    b. Starting at the 3' end, match each nucleotide in the strand to a complementary nucleotide.
    c. Work one at a time and in order, just like it happens in the cell
    d. As you create a pair, connect its phosphate to the sugar of the one added just before it, working down the chain in order.
4. Repeat for the other side of the original DNA molecule.

Q11. What enzyme in DNA replication are you acting as when you are making a copy of the original strand of DNA nucleotides?

Q12. The old DNA is read from 3' to 5'. In what direction is the new strand of DNA nucleotides made during DNA replication?

Q13. Compare the sequence of bases on the two strands within one of the DNA molecules produced by DNA replication. (Comparing left side and right side)

Q14. Compare the sequence of bases between the two separate molecules of DNA resulting from replication. (Comparing the two different puzzles)

Q15. Use what you learned from the puzzle to explain why DNA replication is considered semi-conservative.

Exercise 4.2. Modeling the Cell Cycle

DNA replication occurs during the **synthesis (S) phase** of the cell cycle. The hybrid strands of DNA produced by DNA replication are held together by a mass of proteins called the **centromere**. The goal of **cell division** is to separate these two strands of DNA.

A. Building Chromosomes

A **chromosome** is a unit of DNA and its associated proteins located in a cell. Before DNA replication, chromosomes contain a single strand of DNA and are called **unduplicated chromosomes**. After DNA replication, chromosomes contain two strands of identical DNA (same genes, same DNA sequence) held together by proteins located within the centromere and are called **duplicated chromosomes**. In this exercise, you will model the cell cycle in a cell with *one copy of each of 3 different types of chromosomes*.

Q16. How do different types of chromosomes differ from each other?

Procedure
1. Obtain a ping pong ball-sized piece of modeling clay.
2. Divide the ball in half and put one half aside for now.
3. Use the remaining modeling clay to make 3 "snakes" of different sizes (ranging from 2 to 4 inches). All of the "snakes" should be the same color.

Note: Each "snake" represents a chromosome, a piece of DNA containing many genes.

Q17. Are the chromosomes you built duplicated or unduplicated?

Q18. The chromosomes you built are all different sizes. What does this difference in size represent?

B. Model Interphase

Procedure

Interphase G_1
The cell grows and performs its normal functions.
1. Use a piece of chalk to draw a big circle on your desktop (this represents a cell).
2. Draw a circle inside this circle to represent the nucleus. The circles should be big enough to fit your chromosomes inside!
3. Place the unduplicated chromosomes in the nucleus of the cell.

Interphase S (synthesis)
DNA replication occurs.

4. Use the modeling clay that you set aside earlier to make a second matching snake for each of the chromosomes already in your cell.
5. Attach the new snake to the unduplicated chromosomes in the cell (forming an "X").

The structure you have created is still called a chromosome, but this duplicated chromosome contains two identical sister chromatids.

Q19. How many pieces of DNA do you now have in your cell?

Q20. How many duplicated chromosomes does your cell have after the S phase?

Q21. Draw a diagram of a duplicated chromosome in the margins. Label the centromeres and the sister chromatids.

Interphase G_2
The cell grows and prepares for mitosis by synthesizing important proteins and replicating some membrane-bound organelles. Imagine it happening!

C. Model Mitosis
Mitosis is a type of cell division that is used to separate the sister chromatids in a duplicated chromosome into two unduplicated chromosomes. Mitosis is divided into four stages based on the arrangement of the chromosomes in the cell.

Procedure

Prophase
1. Place the duplicated chromosomes randomly in the center of the cell.
2. Erase the nuclear envelope.
3. Squish your chromosomes up to model DNA condensation.

4. Use chalk to draw spindle fibers from the edge of the cell to each side of each duplicated chromosome.

Metaphase
5. Line duplicated chromosomes up end to end in the center of the cell.
6. Redraw the spindle fibers so that they connect from each side of the cell to each side of each duplicated chromosome.

Anaphase
7. Pull the sister chromatids of each duplicated chromosome apart and place the resulting unduplicated chromosomes about 3 inches apart from each other.
8. Erase the spindle fibers between the separated sister chromatids.

Telophase
9. Move the unduplicated chromosomes to opposite ends of the cell.
10. Erase all spindle fibers.
11. Draw a new nucleus around each set of chromosomes.
12. Lengthen your chromosomes to model DNA decondensing.

Q22. During this process, when did each duplicated chromosome become two unduplicated chromosomes?

D. Cytokinesis

Cytokinesis, cytoplasmic division, is the last phase of the cell cycle. During this phase, the original mother cell is physically divided into two independent daughter cells. This process often occurs at the same time as telophase.

Procedure
1. Draw a line that separates your individual cell into two cells, each with its own nucleus.

Q23. How many chromosomes are in each cell after mitosis and cytokinesis?

Q24. Compare the structure and DNA of the daughter cells you produced to the parent cell that you started with.

And that's the cell cycle. Try it one more time and then demonstrate the process to your instructor.

Get your instructor's initials before moving on: _____

Exercise 4.3. Identifying Cells in Mitosis

In this exercise you will be using the microscope to identify cells in various stages of cell division in onion root tips. Since the tip of a root is a region of rapid plant growth, the cells are dividing rapidly. This allows us to see cells of all the different stages of mitosis in one place. While each of your slides should have cells that show all the stages of cell division, any one cell is frozen in the stage it is in forever—think of each of the cells as a snapshot of one particular time in the cell cycle. Your job will be to identify cells in each of the stages of mitosis.

Each student should work **individually** on this exercise.

Procedure

1. Place a slide of Onion (*Allium*) root tip on the stage of the microscope and adjust until the tip (see figure 4.5) is in focus with the 40× objective.
2. Find a region where cells with condensed chromosomes are visible.
3. Identify the nucleus of the cell. This structure is usually stained pink or purple on our slides.
4. Use the models and posters provided in class to identify cells in each stage of the cell cycle (interphase, cell

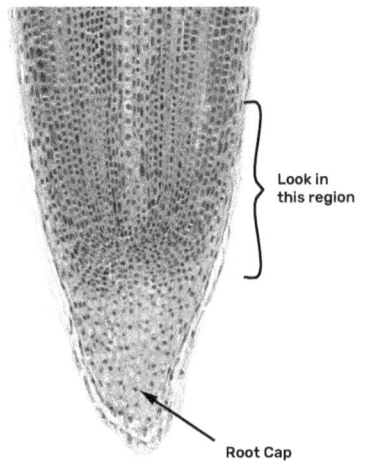

Figure 4.5. Micrograph of an Allium Root Tip. *Look in the region indicated by the bracket to find cells undergoing mitosis.*

Look in this region

Root Cap

66 Cell Biology, Genetics, and Evolution

division, and cytokinesis. Find a cell in each step of mitosis (prophase, metaphase, anaphase, and telophase.
5. Sketch your findings in figures 4.6–4.11. Label any structures you can identify. Be sure to include total magnification and any observations or comments you may have.

> *Skills to Remember*
>
> - Remember to always start on the lowest power
> - Do not switch to the next objective until the image is in sharp focus.
> - Use only the fine focus adjustment knob to sharpen the image at medium and high magnification to avoid cracking the slide.

Figure 4.6. Sketch of an Allium Root Tip Cell in Interphase

Total Magnification: _____

Figure 4.7. Sketch of an Allium Root Tip Cell in Prophase

Total Magnification: _____

Figure 4.8. Sketch of an Allium Root Tip Cell in Metaphase

Total Magnification: _____

Figure 4.9. Sketch of an Allium Root Tip Cell in Anaphase

Total Magnification: _____

DNA Structure, Replication, and Mitosis 67

Figure 4.10. Sketch of an Allium Root Tip Cell in Telophase

Figure 4.11. Sketch of an Allium Root Tip Cell in Cytokinesis

Total Magnification: _____

Total Magnification: _____

Exercise 4.4. Estimating the Time Spent in Each Stage of the Cell Cycle

For this exercise, you will work with your entire class to count cells in the onion root tip in each stage of the cell cycle. Since the cells in the root tip divide randomly, we can use the number of cells at each stage to determine which stage is the longest. For instance, if cytokinesis is really fast, the odds of "catching" a cell in that stage are really low, so we shouldn't find very many cells in that stage.

Procedure

1. For this exercise it will help to **work in teams of two**.
    a. Focus on the region of the root tip containing dividing cells.
    b. One student should look in the microscope while the other acts as a recorder.
2. Each student should identify the phase of cell division of **25 random cells** within the field of view without counting any cell twice. When you count, say the name of the stage aloud so that the recorder can mark it down in table 4.2.
3. After identifying 25 cells, **trade jobs** with your lab partner. Your lab partner should identify the stages of **25 cells in a different part of the slide** while you record the stages.
4. Record your group's data on the class data table provided by your instructor. **Wait until all groups have posted their data before proceeding to the next step.**
5. Calculate the percentage of time spent in each phase by counting the total number of cells in each phase and dividing by the total number of cells you counted. Record the percentage in the appropriate row of table 4.2.

6. In an onion root tip, the entire cell cycle takes about 16 hours. To determine the time spent in each phase of the cell cycle, multiply the percentage of cells in each phase by the total time of the cell cycle (16 hours). Record your answer in table 4.3.

Table 4.2. Number of Cells in Each Stage of the Cell Cycle

Stage	Student 1	Student 2	Your Group Total	Class Total
Interphase				
Prophase				
Metaphase				
Anaphase				
Telophase				
Total Cells counted:				

Table 4.3. Estimate of Time Spent in Each Phase of the Cell Cycle

Phase	Interphase	Prophase	Metaphase	Anaphase	Telophase	Total
% Cells in each phase						100%
Time estimate						16 hours

Q25. Why do we add all of the class data together before doing our calculations? Do you trust the count that other students made? Why or why not?

Q26. Based on your calculations, what stage of the cell cycle is the longest? Provide a hypothesis as to why this is the case.

Q27. Based on your calculations, what stage of mitosis is the longest? Provide a hypothesis as to why this is the case.

Name: _____ Lab Time: _____ Due: _____

4 | Applying What You've Learned in the DNA Structure, Replication, and Mitosis Lab

1. Scientists use the term "anti-parallel" to describe the two strands of nucleotides that form a DNA molecule. Consider the meaning of the terms "anti-" and "-parallel" and to explain how the term antiparallel applies to DNA.

2. The connections between the sugar and the phosphate and between the sugar and the base of a nucleotide fit together like "lock and key" in this puzzle. The bases, however, just slide together (and apart) easily. How does this arrangement in the puzzle reflect the actual properties of these types of bonds in a DNA molecule?

3. Explain and draw the differences between an unduplicated chromosome and a duplicated chromosome.

4. At what stage of the cell cycle and/or mitosis does DNA replicate? During what stage(s) of the cell cycle and/or mitosis would you find the chromosomes in a duplicated state?

5. What happens to the number of chromosomes in a cell during mitosis? For example, if a cell enters mitosis with 8 duplicated chromosomes, how many chromosomes are found in each daughter cell produced by mitosis?

Name: _____ Lab Time: _____ Due: _____

5 | Gene Expression and Mutations Pre-lab

1. Differentiate between the processes of transcription and translation.

2. Complete the diagram in figure 5.1 (on the next page) of the translation machinery of a cell as follows:
   - ❏ Label the ribosome, the mRNA, and one tRNA.
   - ❏ Indicate one codon and one anticodon.
   - ❏ Add the appropriate letters to the anticodons.
   - ❏ Use the genetic code diagram (figure 5.2) to label of each amino acids.

   Hint: The genetic code in figure 5.2 (p. 87) is based on mRNA codons rather than tRNA anticodons.

Figure 5.1. Diagram of a Ribosome Performing Translation. *Made with Biorender*

3. We learned that mutations occur as a result of mismatch errors, but is that the only way they happen?

4. Each of the genes listed in table 5.1 underwent a single-point mutation. Use the information provided to complete table 5.1. Use the final column if you have an idea of what might have caused the changes listed.

Table 5.1.

Gene	Sequence of Polypeptide	Length of Polypeptide Produced	Did a Frameshift Occur?	Type of Point Mutation (deletion, insertion or substitution)
1	Only one change in the amino acid sequence	No change in polypeptide length		
2	Starts out identical but differs after the site of mutation	Longer than the original polypeptide		
3	No change in the sequence	No change in polypeptide length		
4	Starts out identical but differs after the site of mutation	Shorter than the original polypeptide		
5	No change in sequence, but most of the polypeptide is missing	Very short polypeptide-no amino acids exist after the site of mutation		

Watch the video *The Evolution of Lactose Tolerance* (15 min) by scaning the QR code or visiting https://go.chemeketa.edu/lactose and answer the following questions:

What types of food contain the sugar lactose?

- ❑ Milk and dairy products
- ❑ Meat
- ❑ Grains
- ❑ Fats

What simple sugars are produced when the enzyme **lactase** breaks down the sugar **lactose**? *Check all that apply.*

- ❑ Glucose
- ❑ Galactose
- ❑ Sucrose
- ❑ Maltose

At what age can most humans digest lactose?

- ❏ Infancy & Early Childhood (ages 1–8 years old)
- ❏ Late Childhood & Adolescence (8–19 years old)
- ❏ Adulthood (over 19 years)

What does lactase persistence mean? *Check all that apply*

- ❏ The ability of an adult to produce the lactase enzyme
- ❏ The ability of an adult to digest lactose
- ❏ The ability of an adult to produce lactose

Lactase persistence is caused by a mutation. Where is that mutation located?

- ❏ In the gene that contains the instructions to build the lactase enzyme
- ❏ In non-coding DNA that regulates the lactase gene
- ❏ In the mRNA used to build the lactase protein
- ❏ In the genetic code of lactase persistent people

Name: _____ Lab Time: _____ Due: _____

5 | Gene Expression and Mutations

By the end of this lab, you should be able to:

- ❏ Convert a sequence of DNA to a sequence of RNA.
- ❏ Use the genetic code to build a short protein from a sequence of RNA.
- ❏ Describe the importance of protein structure in regard to function.
- ❏ Provide an example of the regulation of gene expression.

Exercise 5.1. How Do Cells Use DNA?

DNA is considered the "blueprint of life," but how does this stringy molecule determine the physical and metabolic features (**phenotype**) of an organism? This process, known as **gene expression**, involves three steps:

1. **Transcription**: synthesizing molecule of RNA from a specific gene in a strand of DNA
2. **Translation**: synthesizing a chain of amino acids from a molecule of RNA
3. **Protein folding and function**

A. Complete the DNA Molecule

DNA is made of two strands of DNA nucleotides held together by weak hydrogen bonds between complementary bases.

Procedure
1. Find the two **red** laminated DNA sections in your Protein Synthesis Kit.
2. Align the unlettered strand of DNA nucleotides to the lettered strand.
3. Use an erasable marker to indicate the order of nucleotides (A, T, G, or C) in the unlettered strand.
4. Label the 3'- and 5'- ends of each strand of nucleotides with the marker.

Q1. How do you know which base to put where on the unlettered strand of DNA nucleotides?

B. Transcription

During **transcription**, enzymes unzip the DNA for a particular gene, and then the enzyme RNA polymerase creates a complementary RNA copy of ONE SIDE of the DNA molecule.

Procedure

1. Unzip the DNA molecule you prepared in part A by separating the two strands of the double helix (red). *The pre-printed strand of nucleotides will act as the template strand for this exercise.*
2. Place the green laminated RNA strand between the two DNA strands with the bases (teeth) facing the template strand. *The teeth of the DNA should line up with the teeth of one strand of the RNA.*
3. Label the 3′ and 5′ ends of the RNA molecule with a dry-erase marker.
4. Starting at the 5′ end of the RNA molecule, use a dry-erase marker to indicate the order of nucleotides (A, U, G, or C) in the unlettered strand. *Adenine in DNA is complementary to Uracil in RNA.*
5. Move the RNA molecule out from between the two DNA nucleotide strands and place it on the side of your workspace.
6. Move the two DNA strands back together to signify the re-zipping of the DNA following transcription.

Q2. RNA polymerase makes RNA from 5′ to 3′. In what direction does RNA polymerase read the template strand of the DNA molecule?

Q3. How does the RNA molecule produced by transcription differ from the DNA molecule that it was made from?

C. Translation

The process of **translation** uses the information stored in the order of nucleotides in a messenger RNA molecule to form a specific sequence of amino acids. During translation, a ribosome "reads" the sequence of nucleotides in the RNA, three nucleotides (one **codon**) at a time. Each codon corresponds to a specific amino acid in the genetic code (see table 1 in your lab manual). *Unlike DNA replication and transcription, the ribosome reads the RNA in the 5' -> 3' direction.*

Procedure

1. Move your DNA molecule to the side of your workspace and place the mRNA molecule in the center.
2. The nucleotides in the RNA molecule are grouped into sets of three (codons). List the 5 codons in your mRNA (from 5' to 3') here:

Q4. Codons in mRNA: _____ - _____ - _____ - _____ - _____

3. During translation, tRNA molecules enter the ribosome, your workspace, and pair up with each codon. Each tRNA has an exposed anticodon that is complementary to a specific codon. List the 5 anticodons needed for translation of your mRNA here:

Q5. Anticodons corresponding to codons for translation:

4. The codons in the mRNA correspond to specific amino acids. Use the genetic code to determine what amino acids are present in the protein you are creating.

Q6. Amino acids in protein: _____ - _____ - _____ - _____ - _____

> **Double Check:** Did you use the codon or the anticodon to determine which amino acid to use? Our genetic code (see figure 5.2) is based on codons (in the green mRNA molecule). You will get a completely different polypeptide if you use the anticodon in the blue tRNA molecule.

5. Find all the **blue laminated tRNA molecules** in your kit. Use an erasable marker to add the appropriate anticodons to the tRNAs
6. Find all the **yellow laminated amino acids** in your kit. Use an erasable marker to add the appropriate amino acid abbreviations to the yellow circles.
7. Connect the appropriate amino acids (yellow balls) to the "matching" tRNA using the Velcro dots.
8. Starting at the 5'-end, find the tRNA that matches the codon of the first codon in the mRNA and place the anticodon adjacent to the mRNA codon.
9. Find the tRNA that matches the second codon in the mRNA and align its anticodon to the mRNA codon.
10. When you have two adjacent amino acids, use the Velcro dots to simulate the formation of a covalent bond between them.
11. Release the tRNAs from the first codon (leave the amino acid behind. Move down one codon. Continue the process until all codons have been read.

Did you get the right sequence?
Obtain your instructor's initials before moving on: _____

Q7. Each team at your lab bench received a different DNA sequence. How did this difference affect the sequence of the polypeptide produced by transcription and translation?

Q8. What tells the ribosome where along the length of an mRNA molecule to start translation?

Q9. What tells the ribosome where along the length of an mRNA molecule to stop translation?

12. **Clean Up**: Erase the writing from all pieces of DNA, RNA, and protein before returning your kit to your instructor.

Exercise 5.2. How Do Proteins Get Their Structure?

The final step in gene expression involves the actual function of the protein produced by translation. The process of translation produces a long chain of amino acids known as a **polypeptide**. The amino acids that make up that chain all have specific physical properties that determine how they interact with each other. While the amino acids are held together in a chain, the attractions between some amino acids and the repulsions of others cause the change to fold up into a three-dimensional shape. That specific shape determines the protein's function.

In this exercise, we will examine the process of protein folding using beads and pipe cleaners to build a subunit of the protein hemoglobin. Hemoglobin is a protein produced and carried by your red blood cells and functions to increase the amount of oxygen gas (O2) that can be carried by your blood throughout your body. A key feature in the structure of hemoglobin is a tiny "pocket" capable of holding a single molecule of O2.

Procedure

1. Obtain the "Transcription-Translation Tables" from your instructor. Follow the procedure to complete this worksheet.

The gene for hemoglobin contains thousands of base pairs. For the sake of simplicity, we will focus on the small region of this gene that encodes the "pocket" that binds O2. The sequence of that gene region is shown at the top of the Transcription-Translation Table. The template (sense) strand of the gene is the one shown on top on the worksheet. It has been transferred to the table for you.

2. Use an erasable marker to transcribe the DNA sequence in the Transcription-Translation Table to the RNA sequence using the spaces provided below the sequence.
3. Use the color-coded copy of the genetic code provided for this activity to translate the mRNA into a polypeptide by listing the appropriate amino acids in the Transcription-Translation Table. Use an erasable marker to record the amino acid abbreviation in the space provided in the table.
    a. During translation, mRNA is read from 5' to 3'.
    b. Remember that this is only part of the gene. The start codon and stop codon are not present in this sequence.

Q10. What level of protein structure is indicated by the sequence of amino acids you just produced?

4. Using the color-coded genetic code determine the color of bead that should be used to represent each amino acid in the polypeptide produced by transcription and translation of the provided hemoglobin gene sequence. Indicate the color of the bead needed in the appropriate cells of the **Transcription-Translation Table**.

To demonstrate protein folding, we will build a model of the hemoglobin protein with a pipe cleaner and beads. The structure of the protein will be based on properties assigned to the beads based on the amino acids they represent.

> The color-coded copy of the genetic code emphasizes the interaction of each amino acid to the presence of water.
> - ❑ Red amino acids are considered hydrophobic.
>   - » Hydrophobic amino acids are water-loving/water-hating (circle one).
> - ❑ Blue amino acids are considered hydrophilic.
>   - » Hydrophilic amino acids are water-loving/water-hating (circle one).
> - ❑ Yellow amino acids have an intermediate interaction with water.

5. Obtain a white pipe cleaner and a selection of colored beads
6. Bend the ends of the pipe cleaner over the beads on each end.
7. Space the remaining beads evenly along the pipe cleaner. The beads should be approximately a thumb-width apart.
8. To demonstrate secondary protein structure, bend the pipe cleaner slightly between each pair of beads (making a zigzag strand). Recall that the secondary structure includes either spiral helices or pleated sheets.
9. To demonstrate tertiary protein structure, bend the pipe cleaner to form a three-dimensional "blob." How you bend the strand is not random. Recall that the amino acids

have different levels of interaction with water. Proteins evolved in the aquatic environment of the cell. As a result, hydrophobic amino acids tend to cluster together deep inside a protein, while hydrophilic amino acids form a protective shell on the outside.

As you form your protein, bend the pipe cleaner such that all the hydrophobic amino acids (red beads) are clustered together at the very center while the hydrophilic amino acids (blue beads) are on the outside. The yellow intermediate amino acids should end up somewhere in between.

10. Does your three-dimensional hemoglobin protein have a small pocket? That is where the O_2 fits. The red pompom in your kit represents O_2. Fit the pompom into the pocket of the protein you created.
11. Functional hemoglobin requires four hemoglobin proteins. Fit your hemoglobin protein together with those of three other teams in your class to produce a functional hemoglobin complex.

Q11. What level of protein structure does this complex represent?

Exercise 5.3. Where Does a Protein Get Its Structure?*

The three-dimensional structure of hemoglobin is key to its ability to bind and hold O_2 in our blood. Even small changes in the sequence of the hemoglobin gene can change this structure and make it harder for O_2 binding to occur. One such mutation is known to be the cause of sickle cell anemia, a genetically inherited blood disorder associated with a variety of chronic health conditions.

> **Case study:** In 2007, NFL player Ryan Clark was rushed to the hospital during a game in Denver, Colorado, with severe pain in his side. His spleen and gallbladder were removed as a result of complications from carrying one copy of a mutant form of the hemoglobin gene. Performing strenuous activity at high altitude exacerbated his symptoms.

Sickle cell anemia is a disease caused by a mutation in the hemoglobin gene. A change in just one specific nucleotide in the gene for this protein causes a change in its amino acid sequence. That change alters the three-dimensional protein structure and decreases the ability of hemoglobin to bind oxygen gas. The mutant protein is also "sticky" and tends to form

---

* Based on Williams et al (2014) Hook Students with the Super Bowl and a Protein Modeling Activity to Teach Genetic Concepts, *J. Microbiology and Biology Education*, 15(1): 41-42.

clumps in the red blood cells, especially when O_2 levels are low. The result is some or all of the red blood cells in a sickle cell anemia patient are sickle ("C-") shaped. Sickle-shaped cells do not flow as smoothly through the capillaries as normal red blood cells and can form clumps that block blood flow, causing damage to internal organs.

In this exercise, we explore how a change in just one nucleotide in a gene can cause a disease like sickle cell anemia by building the mutant hemoglobin protein using pipe cleaners and beads.

Procedure

1. To continue with this lab, return to the "Transcription-Translation Tables" used in Exercise 5.2.
2. Compare the DNA sequence of the mutated gene to the DNA sequence of the normal hemoglobin gene (in Exercise 5.3). Circle or highlight the base pair that is different in the mutant gene.
3. Use the process of transcription to build an RNA copy of the hemoglobin gene sequence provided. Record the RNA sequence in the appropriate cells of the Transcription-Translation Table.
4. Compare the RNA sequence of the mutated gene to the RNA sequence of the normal hemoglobin gene (Exercise 5.2). Circle or highlight the mRNA codon that is different in the mutant gene.
5. Use the color-coded copy of the genetic code provided for this activity to translate the mRNA into a polypeptide by listing the appropriate amino acids in the Transcription-Translation Table.
6. Compare the amino acid sequence of the mutated polypeptide to the amino acid sequence of the normal hemoglobin polypeptide. Circle or highlight the amino acid that is different.

Q12. How does the amino acid sequence of the mutated polypeptide differ from that of the normal hemoglobin polypeptide with regard to the interaction of the molecule with water?

Q13. How do you think this change will affect the three-dimensional structure of the hemoglobin protein?

7. Using the color-coded genetic code determine the color of bead that should be used to represent each amino acid in the polypeptide produced by transcription and translation of the mutant version of the hemoglobin gene sequence. Indicate the color of the bead needed in the appropriate cells of the Transcription-Translation Table.
8. Once you have determined the bead colors for all the amino acids, thread them onto the green pipe cleaner in order. The beads should be approximately a thumb-width apart.
9. Bend the pipe cleaner to form a three-dimensional "blob." Remember that protein folding is not random. Return to Exercise 5.2 to review the rules.

Q14. Compare the overall structure of the normal hemoglobin you built in Exercise 5.2 with the mutant hemoglobin you built in this exercise. Does the overall shape of the protein differ?

Q15. Does the mutant hemoglobin protein have a pocket (or at least the same sized pocket) as the normal hemoglobin protein for holding O_2?

You have two copies of almost every gene in your body. A person with sickle cell anemia has two copies of the mutated version of the hemoglobin gene. Ryan Clark has one "good" copy of the hemoglobin gene and one mutant version (a condition known as sickle cell trait).

Q16. Why didn't Ryan Clark exhibit symptoms of sickle cell anemia under normal conditions?

Q17. Why did Ryan Clark's body react negatively during a football game played in Denver, Colorado, the "Mile-high" city?

Figure 5.2. The Genetic Code. mRNA codons for amino acids are used in protein synthesis. The color bars on the outside of the circle indicate the chemical properties of the associated amino acid. Red amino acids are hydrophobic, blue amino acids are hydrophilic, and yellow amino acids are intermediate.

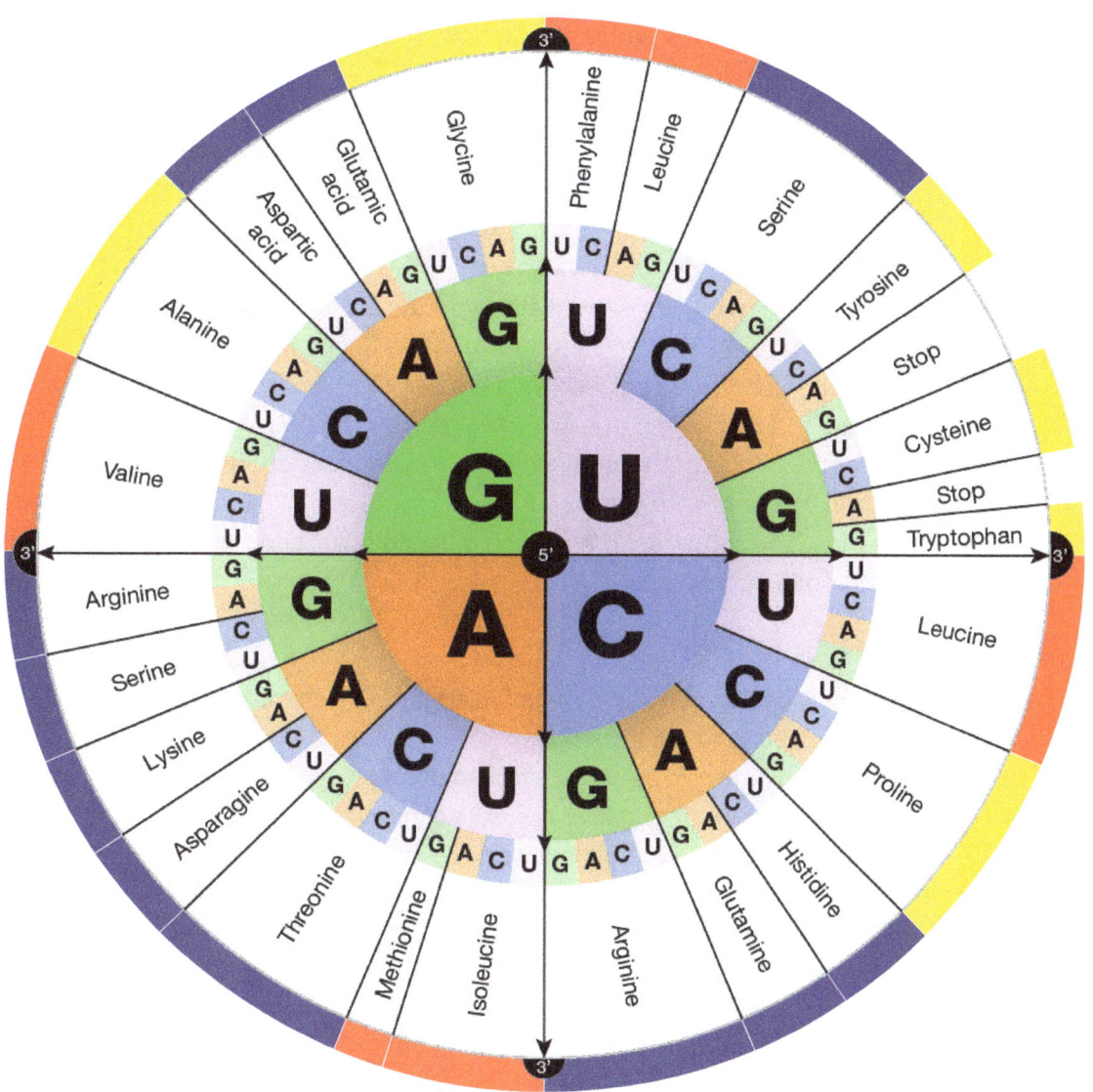

Exercise 5.4. How are Genes Regulated?**

In the video *The Evolution of Lactose Tolerance*, we learned that lactase is an enzyme that breaks down the sugar lactose. Not all human cells produce the lactase enzyme—it is only made in the cells of the small intestine. Lactose is too big of a sugar for those cells to absorb, so the enzyme breaks it into smaller sugars—glucose and galactose, that intestinal cells can absorb. In humans, the gene for lactose is regulated, which means that it is not usually made

---

** Modified from "Milk–How Sweet Is It?" by www.biointeractive.com.

throughout the life of a human. Gene regulation helps cells conserve energy by only making the proteins needed at the time they are needed. Ancient humans only consumed lactose while breastfeeding. As a result, the gene for lactase usually remains inactive after early childhood. Some humans, however, have evolved the condition of lactase persistence, which allows their intestinal cells to continue producing the enzyme into adulthood. As a result, lactase-persistent humans are able to consume dairy products as adults.

In this activity, your group will test for lactase persistence by adding intestinal solutions from two patients to milk and measuring changes in the amount of glucose in the milk over time.

How to Use a Glucose Test Strip
1. Dip the test area of the glucose test strip into the patient 1 beaker and immediately remove it.
2. Lay the test strip on the laminated Glucose Test Strip Chart in the slot labeled Patient 1, 0 min.
3. Wait 30 seconds. This test is time sensitive. Waiting longer than 30 seconds will make your results less accurate.
4. Compare the color of the test area to the glucose color chart to determine the glucose concentration.

Procedure
1. Label 4 beakers as "positive control," "negative control," "patient 1," and "patient 2."
2. Add the following solutions to each beakers:
    a. Positive Control: 5 ml of positive control solution
    b. Negative Control: 5 ml of negative control solution
    c. Patient 1: 5 ml of patient 1 intestinal fluid
    d. Patient 2: 5 ml of patient 2 intestinal fluid
3. Arrange the beakers above their respective sections of the laminated Glucose Test Strip Cart.
4. Add 5 ml of milk to each of 4 additional beakers
5. Place one milk beaker above each sample beaker.
6. BEFORE adding any milk, use a test strip and the procedure outlined in the "How to Use a Glucose Test Strip" box to determine the baseline level of glucose in each sample. Record the results in the 0 min (baseline) column of table 5.2.
7. Answer the following questions before moving on.

The chemical reaction we are investigating is:

$$\text{Lactose} \xrightarrow{\textit{Lactase}} \text{Glucose} + \text{Galactose}$$

Q18. What is the source of lactose in this experiment?

Q19. What is the enzyme that digests lactose?

Q20. What chemical are we testing for with the test strips?

Q21. What will the positive control for this experiment tell us?

Q22. What part of the chemical reaction is not used in the negative control? Why is this considered a negative control?

Table 5.2. Results of Testing for Lactase Activity in Patient Intestinal Samples

Samples	Glucose (mg/dL) present at...		
	0 min (baseline)	2 min	7 min
Negative Control			
Positive Control			
Patient 1			
Patient 2			

8. Test for lactase activity, using a new test strip for each sample and time point.
    a. Pour the milk into the negative control beaker and stir.
    b. Start the timer immediately. You will retest the liquid in 2 min and 7 min.
    c. At 2 min, test the solution in the negative control beaker using the procedure provided in the box at the start of this activity. Record the data in the 2 min column of table 5.3. Repeat with remaining samples.
    d. At 7 min, test the solution in the negative control beaker using the procedure provided in the box at the start of this activity. Record the data in the 7 min column of table 5.3. Repeat with remaining samples.
9. Build a bar graph of the data collected in the space provided in figure 5.3.
    a. Give the graph a title.
    b. Label the X-axis, Y-axis, and a legend.
    c. The columns should represent the glucose level for each sample 0 min, 2 min, and 7 min.

Figure 5.3. Bar Graph of Collected Data for Exercise 5.4

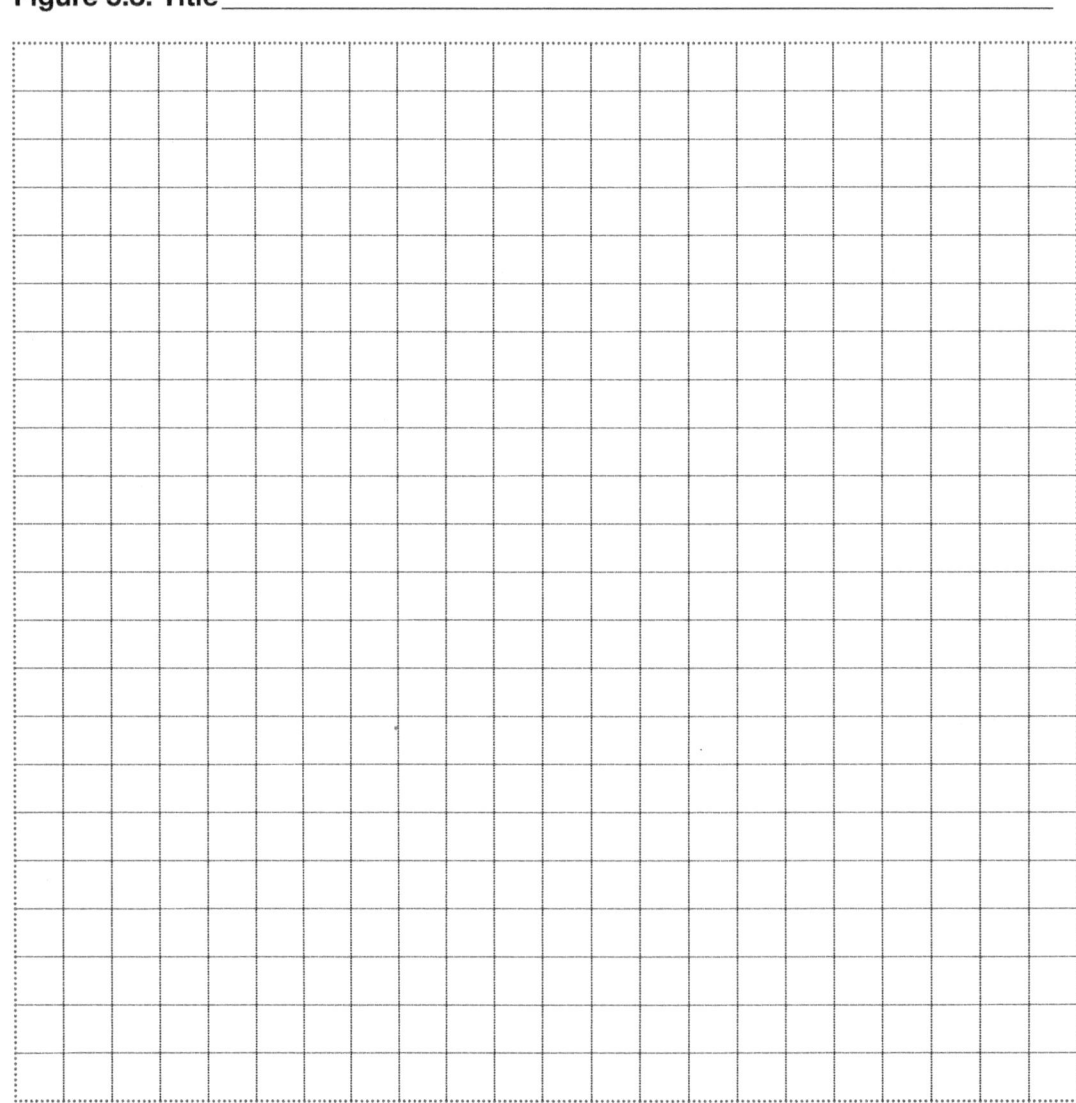

Figure 5.3. Title _____

90 Cell Biology, Genetics, and Evolution

Q23. Are either of the patients lactase-persistent? Describe the evidence that supports this claim?

Q24. Are either of the patients lactase-non persistent (lactose intolerant)? Describe the evidence that supports this claim?

Q25. How did the negative control help you interpret the results of this experiment?

Q26. How did the positive control help you interpret the results of this experiment?

Q27. Does being lactose intolerant mean that you do not have the gene for the lactase enzyme? Explain.

Q28. Lactase-persistence and lactose-intolerance are examples of gene regulation. Provide a basic description of how the lactose gene is regulated in humans that exhibit lactose intolerance.

Name: _____ Lab Time: _____ Due: _____

5 | Applying What You've Learned in the Gene Functions and Mutations Lab

1. Biologists use a variety of chemicals to explore the processes of transcription and translation in cells. Suppose a scientist treated a cell with a chemical that inhibits transcription. Would translation be able to occur? Why or why not?

2. Explain how translation differs from transcription.

3. What is the role of the anticodon in the process of translation?

4. Is it possible for a mutation to change the DNA sequence of a gene but not the amino acid sequence of the polypeptide that the gene produces? Use the genetic code to explain your answer.

5. Is it possible for a mutation to change the amino acid sequence of a polypeptide but not the tertiary structure of the protein that the polypeptide folds into? Use what you learned in today's lab to explain your answer.

6. If lactose intolerance did not cause any symptoms, lactose-intolerant individuals could use milk as a source of protein. Explain why it is possible for them to digest the proteins in milk even though they lack the lactase enzyme.

7. Some lactose-intolerant individuals consume lactose-free milk. What would you expect to happen to the glucose levels of a lactose-intolerant individual after consuming lactose-free milk?

8. Sucrose is a disaccharide present in many milk substitutes, such as soy, rice, and almond milk. It is composed of glucose and fructose. Explain why lactose-intolerant individuals may be able to digest sucrose without any problems.

Name: _____ Lab Time: _____ Due: _____

6 Meiosis and Sexual Reproduction Pre-lab

Answer the following questions as you watch "What's the Point of Sex?" (14 mins). Scan the QR code or visit https://go.chemeketa.edu/sex to access the video.

1. How does Muller's Ratchet apply to reproduction?

2. How do organisms that do not perform sexual reproduction avoid Muller's Ratchet?

3. What are the two processes required for sexual reproduction?

4. What are the "downsides" of sexual reproduction?

5. Why is an understanding of meiosis important to understanding the process of sexual reproduction?

6. Draw a duplicated chromosome, a homologous pair of duplicated chromosomes and two unduplicated, nonhomologous chromosomes in the boxes in figure 6.1.

Figure 6.1. Sketches of Different Chromosomal Arrangements

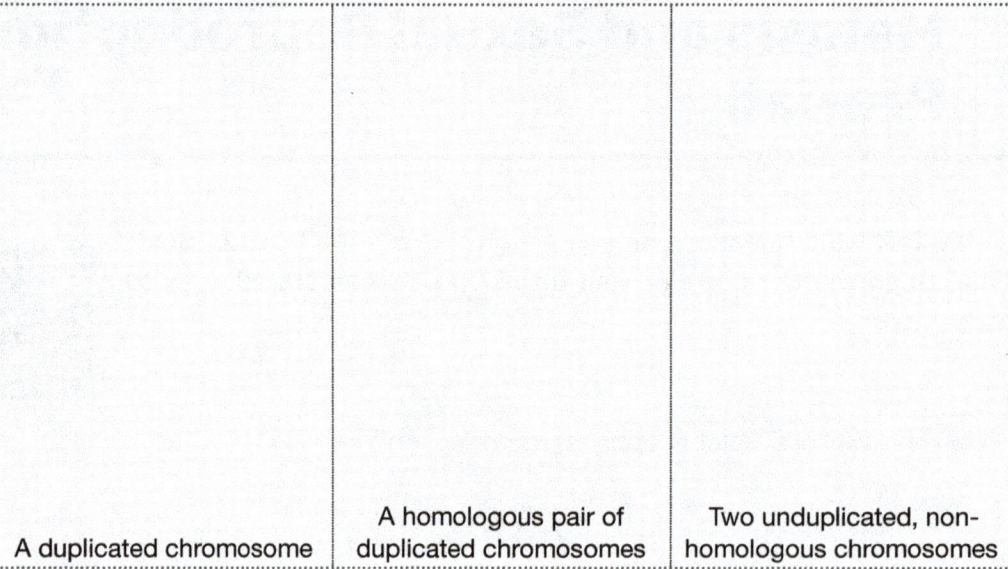

| A duplicated chromosome | A homologous pair of duplicated chromosomes | Two unduplicated, non-homologous chromosomes |

7. Determine whether the events described in table 6.1 are part of mitosis or meiosis (or both) by checking the appropriate boxes in the table.

Table 6.1.

Which process(es)….	Mitosis	Meiosis
performs cytokinesis twice		
ends with 4 cells		
creates new cells		
changes a 2n cell into several 1n cells		
changes a 1n cell into two 1n cells		
includes crossing-over and independent assortment		
requires DNA replication before the process begins		

Name: _____ Lab Time: _____ Due: _____

6 | Meiosis and Sexual Reproduction

By the end of this lab, you should be able to:

- ❑ Differentiate between the two types of cell division.
- ❑ Use chromosome models or drawings to explain the process of meiosis to another student.
- ❑ Explain how crossing over increases genetic diversity.
- ❑ Explain why sexual reproduction is important to inheritance and evolution.

Exercise 6.1. Modeling Meiosis with Clay

In this exercise you will be modeling the movement of chromosomes through the eight phases of meiosis. Recall that we learned that humans have 23 different types of chromosomes and each of your cells (except sperm or egg cells) have two versions of each of these chromosomes for a total of 46 in each cell. For the sake of simplicity, you will model meiosis in a **cell with a diploid number of 4 ($2n = 4$)**.

- ❑ This equation means that the cell you are building has two copies of each of the two types of chromosomes.
- ❑ Each type of chromosome can be either unduplicated or duplicated, depending on the stage of the cell cycle.

Q1. During what stage(s) of the cell cycle are chromosomes found in their duplicated form?

Q2. During what stage(s) of the cell cycle does a diploid cell contain homologous pairs?

Q3. During what stage(s) of the cell cycle does a haploid cell contain homologous pairs?

Q4. Your model cell is diploid and contains a total of four chromosomes.

How many types of chromosomes will be in your cell? _____

How many pairs of homologous chromosomes will be in your cell? _____

A. Chromosome Modeling

Build a total of 4 chromosomes: two long and two short. The two chromosomes of the same size should be different colors

Procedure:
1. Obtain a ping pong ball-sized chunk of each of two colors of modeling clay.
2. Use the clay to build four duplicated chromosomes out of thick "snakes" as follows:
    a. Color 1, long
    b. Color 2, long
    c. Color 1, short
    d. Color 2, short
3. Pile your chromosomes randomly in the center of your workspace (the cell).

Q5. Assign one color of clay to "Mom." Indicate that color here._____

Q6. Assign one color of clay to "Dad." Indicate that color here _____

Q7. Who are your long chromosomes inherited from?
- ❏ both from mom
- ❏ both from dad
- ❏ one from mom and one from dad

Q8. Who are your short chromosomes inherited from?
- ❏ both from mom
- ❏ both from dad
- ❏ one from mom and one from dad

---

98 Cell Biology, Genetics, and Evolution

Q9. Which of the following constitute(s) a homologous pair of chromosomes? *Check all that apply.*
- ❑ Long, blue, duplicated chromosome and long, yellow, duplicated chromosome
- ❑ Long, yellow, unduplicated chromosome and long, blue, unduplicated chromosome
- ❑ Long, blue, duplicated chromosome and short, blue, duplicated chromosome
- ❑ Short, yellow, unduplicated chromosome and long, yellow, unduplicated chromosome

Q10. What stage of interphase does the pile of chromosomes in your workspace represent? How do you know?

Q11. How many times was the DNA in your cell replicated to reach this stage?

B. Meiosis I

The goal of Meiosis I is the separation of homologous pairs. This process reduces the chromosome number of a cell from diploid to haploid. In addition, the chromosomes undergo **crossing over** during prophase I and **independent assortment** occurs during metaphase I.

Procedure

Prophase I
1. Move duplicated chromosomes of the same size (and different colors) together. In this arrangement, the chromosomes are often called **tetrads** because they contain 4 pieces of DNA.

> **Note: Do not perform crossing over at this time.**

Q12. Select the term(s) that apply to one tetrad in your model. *Check all that apply.*
- ❑ Unduplicated
- ❑ Duplicated
- ❑ Homologous Pair
- ❑ Different types of chromosomes

Q13. Which is stronger, the bond between the sister chromatids in your tetrad or the bond between homologous chromosomes in your tetrad?

Metaphase I
2. Move the tetrads into a vertical line at the center of the cell. Demonstrate **independent assortment** by determining how many different arrangements of chromosomes this cell could form.

Q14. How many different arrangements of chromosomes are possible at metaphase I in your model?

Q15. How does independent assortment contribute to the generation of genetic diversity?

Anaphase I
3. Separate the two duplicated chromosomes in each tetrad and move them about 4 inches apart.

Q16. What did you separate at this stage? Homologous pairs or sister chromatids?

Telophase I
4. Move the separate duplicated chromosomes to opposite ends of the cell forming two piles.
5. Draw a line with chalk to divide your workspace in half. This physical division represents cytokinesis. Each side of your workspace represents a cell formed by cytokinesis after meiosis I.

Q17. How many cells are formed by meiosis I?

Q18. How many chromosomes are in each cell?

Q19. Is the resulting cell diploid or haploid?

Q20. How does meiosis I differ from mitosis?

C. Meiosis II

Depending on cell type, there may be a brief break between meiosis I and II. During this break, chromosomes may unwind, but DNA replication will *not* occur again. **Meiosis II** is the stage in which sister chromatids are separated; duplicated chromosomes become unduplicated chromosomes.

Procedure

Prophase II
1. Randomly arrange the duplicated chromosomes in the center of each cell

Metaphase II
2. Align duplicated chromosomes head to tail in a horizontal line in each cell

Q21. How is the arrangement of duplicated chromosomes in metaphase II different from the arrangement of duplicated chromosomes in metaphase I?

Anaphase II
3. Separate the sister chromatids of each duplicated chromosome and move them slightly toward opposite poles (about 4 inches apart).

Telophase II
4. Move the resulting unduplicated chromosomes to opposite sides of each cell.

Cytokinesis
5. Use chalk to make a line between the two clusters of chromosomes in each cell, creating separate cells.

Q22. How many cells does Meiosis II produce?

Q23. Are the resulting cells haploid or diploid?

Q24. How is Meiosis II similar to mitosis?

D. Crossing Over

During **crossing over**, the non-sister chromatids of a homologous pair exchange bits of DNA. Discuss how your model could be modified to include crossing over with your partner.

Procedure
1. Perform parts A and B again, but include crossing over by breaking a small piece of one chromatid in a homologous pair and exchanging it with a small piece of DNA from the chromatid of the other member of the homologous pair.

Q25. Explain how the gametes produced during meiosis with crossing-over are different from the gametes produced during meiosis without crossing-over.

Go through the process a couple more times and then model it for your instructor.

Get your instructor's initials before moving on: _____

Q26. A cell has 3 homologous pairs. Each pair carries a different gene (A, B, or E). Draw all the possible ways that these chromosomes could line up (including letters) in metaphase I in the boxes in figure 6.2.

Figure 6.2. Diagramming Independent Assortment

Exercise 6.2. Meiosis in *Lilium* anthers

Meiosis is difficult to observe in humans because it occurs deep within our bodies, and the cells are tiny. Instead, pollen grains are commonly used to study meiosis because they are easy to find and because the daughter cells stay attached to each other throughout the process. Meiosis turns one cell into four daughter cells, so the stage of the meiosis is easily determined by counting the number of cells (figure 6.3) and observing the arrangement of chromosomes within each cell of the pollen grain.

Figure 6.3. Arrangement of Developing Pollen Grain Cells During Meiosis

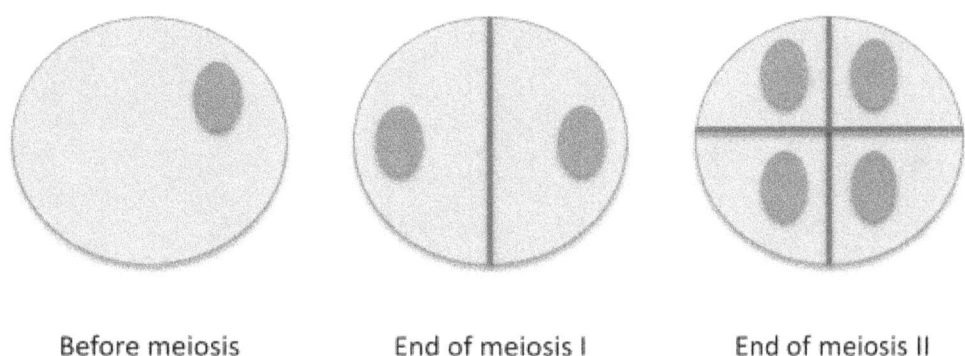

Before meiosis   End of meiosis I   End of meiosis II

Q27. Check the box of the chromosome number that applies to the "Before Meiosis" cell in figure 6.3.
- ❑ 1n (haploid)
- ❑ 2n (diploid)

Q28. Check the box of the chromosome number of the cells in a developing pollen grain that has two cells (figure 6.3, "End of Meiosis I").
- ❑ 1n (haploid)
- ❑ 2n (diploid)

Q29. Check the box of the chromosome number of the cells in a developing pollen grain that has four cells (figure 6.3, "End of Meiosis II")?
- ❑ 1n (haploid)
- ❑ 2n (diploid)

Q30. How does the amount of DNA in each cell of a pollen grain at the end of meiosis I differ from the amount of DNA in each cell of a pollen grain at the end of meiosis II?

Developing pollen grains undergo meiosis I and meiosis II at the same time, unlike the Allium root tips we used to observe mitosis. You will probably see several stages of Meiosis I in one slide, but you will not see cells from meiosis I and meiosis II on the same slide. As a result, **you will need to look at two slides to see the complete process.**

Figure 6.4. *Lilium* anthers. A. Lily flower with a red line through the flower's anther indicating where the sample in panel B comes from. B. Cross section of a lily anther (Kirchoff, 2015). The red box indicates one of 4 regions where you will find developing pollen.

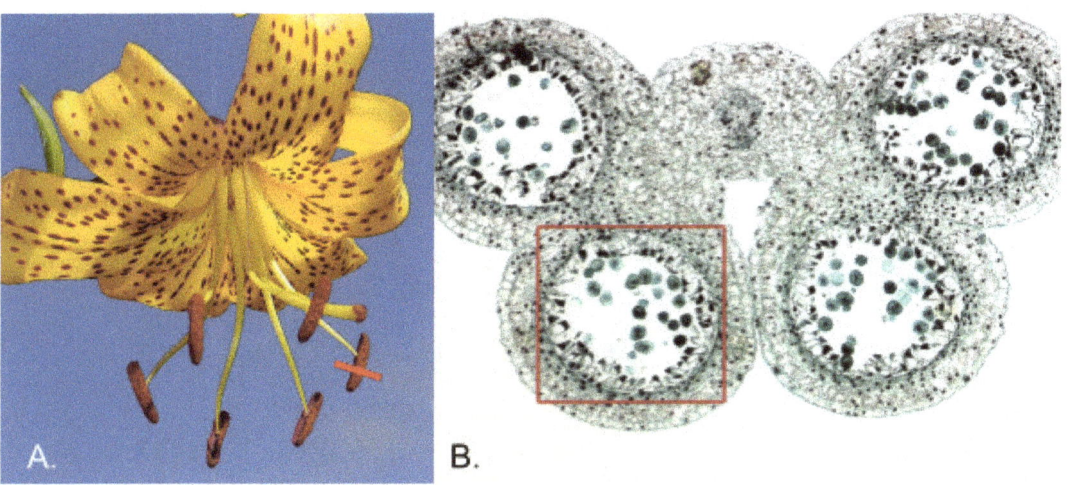

Procedure

1. Use the low-power objective of your microscope to locate the pollen-filled anther on a slide of *Lilium* anther (1st meiotic division).
2. Once centered on the region containing developing pollen (figure 6.4, red box), increase magnification to the highest power.
3. Draw a collection of at least 5 developing pollen grains, including their chromosome arrangement in figure 6.5.
4. If possible, label the stage in meiosis of each developing grain.
5. Repeat this process with the *Lilium* anther (2nd meiotic division slide), draw in figure 6.6.

Figure 6.5. Developing Lily Pollen, First Meiotic Division

Figure 6.6. Developing Lily Pollen, Second Meiotic Division

Q31. Identify and label a cell in figure 6.5 or 6.6 that has not started meiosis yet. How do you know it has not started meiosis? If your sketch doesn't contain this cell, how would you identify a cell that has not started meiosis?

Q32. Do all the cells in figure 6.5 contain homologous pairs? Explain your answer.

Q33. Do the cells in figure 6.5 contain duplicated or unduplicated chromosomes? Explain your answer.

Q34. Do any cells in figure 6.6 contain homologous pairs? Explain your answer.

Q35. Do all the cells in figure 6.6 contain duplicated chromosomes? Explain your answer.

Exercise 6.3. Relating Meiosis to Inheritance

This exercise is designed to help you understand the link between meiosis and genetics (our next topic).

Procedure
1. Complete figure 6.7.
    a. **Draw chromosomes** in the appropriate form (condensed or not; duplicated or unduplicated) in each cell below.
    b. Use two colors, one for parent 1 and one for parent 2.
    c. Distinguish between the two members of a homologous pair in the same parent. Draw one member (maternal) in outline form and the other member (paternal) filled in.
    d. **Label the A gene alleles** (A or a) on each piece of DNA in the drawing.
2. As you work, **answer the questions** in the middle of the diagram.

> **Note:** Since we are only working with one homologous pair, independent assortment and crossing over do not apply.

In genetics, the genotype of an organism is represented by letters.

A single letter represents each copy of a gene that an individual carries.

- The upper and lowercase versions of that letter represent different versions (or **alleles**) of that gene.
- Animals use meiosis to make gametes, which are haploid. As a result, the genotype of a gamete includes one letter for any given gene.

Figure 6.7.

Maternal Germ Cell
Genotype: ____

Complete the diagram by adding chromosomes

Circle the term that applies to each of the following stages

Paternal Germ Cell
Genotype: ____

Before / after S phase
Haploid / Diploid
Duplicated / Unduplicated
Condensed / Decondensed

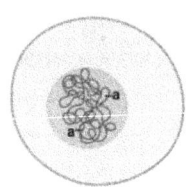

Start of Meiosis
Haploid / Diploid
Duplicated / Unduplicated
Condensed / Decondensed

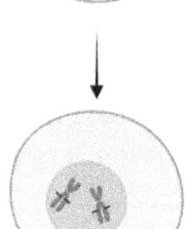

Products of Meiosis I
Haploid / Diploid
Duplicated / Unduplicated

Products of Meiosis II
Haploid / Diploid
Duplicated / Unduplicated

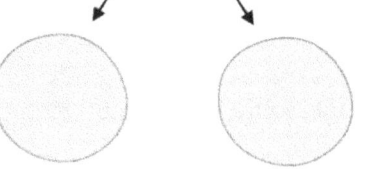

Gametes

Indicate the genotype of each cell produced by meiosis in the space below each gamete

Circle one gamete from each parent to indicate which gametes will undergo fertilization to create a new individual

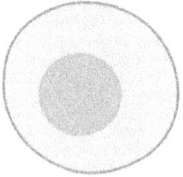

Zygote
Haploid / Diploid
Duplicated / Unduplicated
Condensed / Decondensed

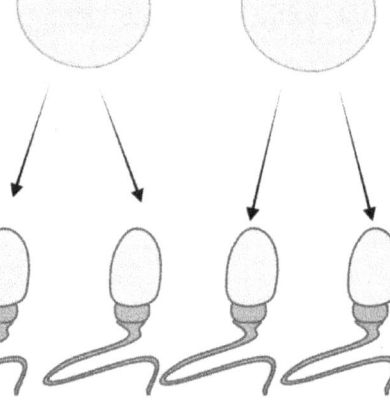

108   Cell Biology, Genetics, and Evolution

Q36. Use this information to complete table 6.2, comparing the genotypes of the individuals and gametes depicted in the diagram on the previous page. If more than one genotype is possible, list all possible genotypes.

Table 6.2.

	Genotype(s)
Parent 1	
Parent 2	
Gametes of parent 1	
Gametes of parent 2	
Zygote	

Q37. Compare your table 6.2 to those of other students in your group and in other groups. Which cells are always the same? Which can differ? What explains the differences?

Q38. Recall that sexual reproduction requires meiosis and nuclear fusion. Where in the diagram on the previous page did these two processes occur?

Q39. In your own words, explain how an understanding of meiosis can help you understand how genes are inherited (genetics).

Name: _____ Lab Time: _____ Due: _____

6 | Applying What You've Learned in the Meiosis & Sexual Reproduction Lab

1. What are the two processes required for sexual reproduction to occur?

2. Differentiate between a pair of sister chromatids and a homologous pair (you may want to include a sketch).

3. During what stage(s) of the cell cycle and/or meiosis are homologous pairs *present* in a cell?

4. During what stage(s) of the cell cycle and/or meiosis are homologous pairs actually found in pairs (tetrads) in a cell?

5. What happens to the number of chromosomes in a cell during meiosis? For example, if a cell enters meiosis with 8 duplicated chromosomes, how many chromosomes are found in each daughter cell produced by meiosis?

6. Describe at least three ways that meiosis and sexual reproduction increase genetic diversity.

Name: _____ Lab Time: _____ Due: _____

7 | Genetics Pre-lab

Answer the following questions while watching the TedEd video "How Mendel's Pea Plants Helped Us Understand Genetics." Scan the QR code or visit http://go.chemeketa.edu/mendelspeaplants to access the video.

1. What are alleles?

2. Why did Mendel call the yellow pea allele dominant?

3. What is the purpose of a Punnett Square?

4. What goes on the side and top of a Punnett Square? (check one)

    ❏ Gamete genotypes
    ❏ Offspring genotypes
    ❏ Gamete phenotypes
    ❏ Offspring phenotypes

5. What do the boxes inside a Punnett square represent?

6. Complete table 7.1 by putting an "X" in the boxes of all terms that definitely apply to the example.

Table 7.1.

Example	Genotype	Phenotype	Homozygous	Heterozygous	Haploid (1n)	Diploid (2n)
g						
Gg						
Green						
White						
gg						

7. How many genes must you be studying for the law of independent assortment to apply?

8. A cell with 3 homologous pairs (each pair carries a different gene: A, B, or E) can create 8 different cells as a result of independent assortment. Draw all 8 possible ways that these chromosomes could line up in metaphase I in figure 7.1. The first two are done for you.

Figure 7.1. Diagramming the Process of Independent Assortment

9. A true-breeding curly-haired sheep is crossed to a true-breeding straight-haired sheep. The sheep produce two curly hair offspring. Which allele is dominant?

Name: _____ Lab Time: _____ Due: _____

7 Genetics

By the end of this lab, you should be able to:

- ❏ Correctly use the terms genotype and phenotype to describe an organism.
- ❏ Predict the genotype and phenotype ratios in a given genetic cross.
- ❏ Explain the results of crosses using the concepts of segregation and independent assortment.

Exercise 7.1. Genetics in Corn

The goal of this activity is to give you the opportunity to walk in Mendel's footsteps. Instead of the garden pea, we will use a variety of corn. Zea mays (maize) exhibits a variety of inheritable phenotypes throughout its life cycle. In today's lab, we will look at phenotypes that are displayed in corn seedlings and in corn seeds.

A. Following the Inheritance of a Single Gene

Observe the flats of F_2 seedlings provided for your class. Answer the following questions:

Q1. What trait do you think we are following in this cross? _____

Q2. What are the two different alleles of this trait? _____

The production of the seeds that produced the individuals you are observing is a bit tricky.
- ❏ The albino allele is fatal as the plants lack chlorophyll and cannot perform photosynthesis.
- ❏ Homozygous white plants cannot be used in a cross.
- ❏ Using more complex genetic breeding strategies, scientists can identify heterozygous plants that represent the F_1 generation.
- ❏ All plants heterozygous for the leaf color gene are green.

Q3. Why is the color of a heterozygous plant important in this cross?

The plants provided for this exercise were produced by crossing two individuals known to be heterozygous for the leaf color gene. Their offspring are the **F$_2$ generation**.

Q4. Complete table 7.2 by assigning a letter to the leaf color gene. Give the dominant allele the capital version of that letter and the recessive allele the lowercase version of the same letter.

Table 7.2. Key to Symbols Used for Leaf Color Alleles in Zea mays

Allele	Symbol
Green Leaves	
Albino Leaves	

Use the letters in table 7.2 to complete figure 7.2.

Figure 7.2. Diagram of the Inheritance of the Leaf Color Gene

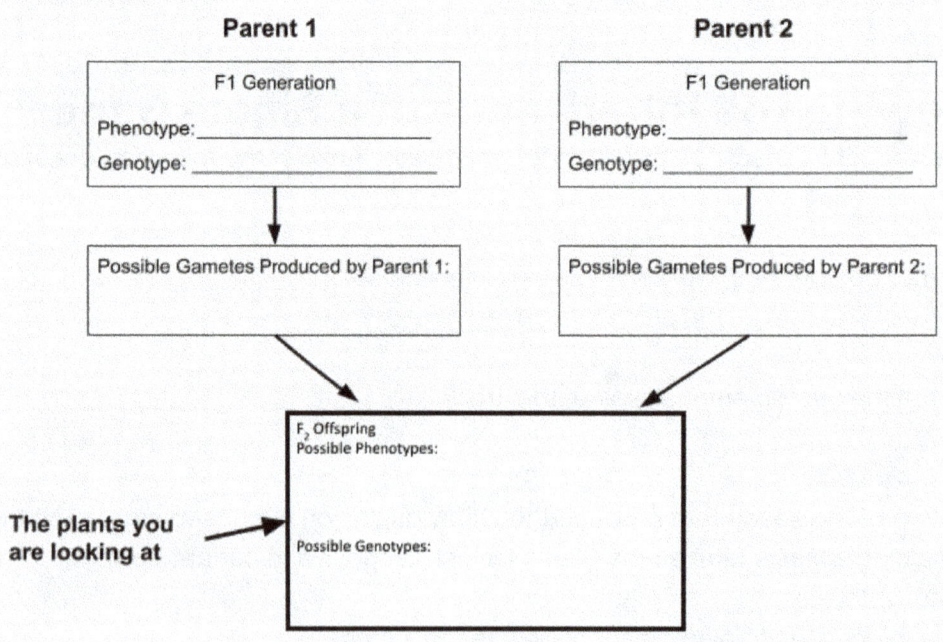

Form a prediction.
- ❏ In genetics, the hypothesis generally states that a particular gene follows a specific pattern of inheritance. In this case, that pattern is the one governed by Mendel's laws.
- ❏ Once we have a hypothesis, we can make a prediction of what the results should be.

> **Hypothesis:** The leaf color gene follows Mendel's laws of segregation.

Q5. What is your prediction? Finish the statement below:

If the leaf color gene follows Mendel's law of segregation, _____% of offspring of a cross between two F1 plants should be green and _____% of the offspring should be albino.

Procedure

1. Count the number of each phenotype present in the corn seedlings.
2. Record your data in table 7.3.
3. Record or obtain data for at least two other flats of seedling.
4. Calculate the % of the population displaying each allele.

Table 7.3. Inheritance Leaf Color in F2 Offspring

Phenotype	Possible Genotype(s)	Number of Seedlings Counted per group			Total	% Total Offspring Counted
		Flat 1	Flat 2	Flat 3		
Green						
Albino						
					Total	100%

Q6. Does your data support your hypothesis? Explain your reasoning.

B. Following the Inheritance of Two Genes

In this exercise, we will examine inheritance by observing corn cobs. The average corn plant produces a single ear of corn with around 500 kernels per ear arranged in discrete rows. Each kernel represents a single new and genetically unique individual. In this example we will focus on how two traits are inherited at once: **kernel color and kernel structure**.

Procedure
1. Examine a parental cross card.
    a. Each side of the **parental cross card** shows the phenotypes of the two different alleles for a particular corn kernel gene, as displayed by members of the P generation (labeled P).
    b. The **image labeled F_1** shows the kernels resulting from the cross between the members of the P generation.
    c. Remember that corn kernels are the offspring of the cross even though they have not germinated into seedlings yet.

Q7. Is the genotype of the P generation of both genes homozygous or heterozygous?

Q8. Is the genotype of the F_1 generation for both genes homozygous or heterozygous?

Q9. Which allele is dominant for each trait? Explain how you know.

2. **Complete table 7.4** by deciding on a letter to represent each gene.
    a. Assign a letter to each of the traits (genes) being studied.
    b. Give the dominant allele the capital version of that letter and the recessive allele the lowercase version of the same letter.
    c. Use the letters in table 7.4 to determine the genotype of the parent plants and gametes each plant produces in figure 7.3.
    d. Remember that your work will be less confusing if everyone in your group uses the same symbols!

Table 7.4. Key to Letters Representing Kernel Color and Structure Alleles in *Zea* mays

Trait: Kernel Color	Symbol	Trait: Kernel Structure	Symbol
Purple		Smooth	
Yellow		Rough	

In this exercise we will look at both of these genes at the same time.
- The **phenotype of an individual**, you should include two traits (kernel color and kernel structure).
- The **genotype of each individual** should contain four letters (AaBB).
- Gametes have **one copy of each allele** and should include two letters (Ab).

3. Use your symbols from table 7.4 and the parental cross card to **complete figure 7.3**.

Figure 7.3. Diagram of the Cross Between the True-Breeding P Generation (two traits)

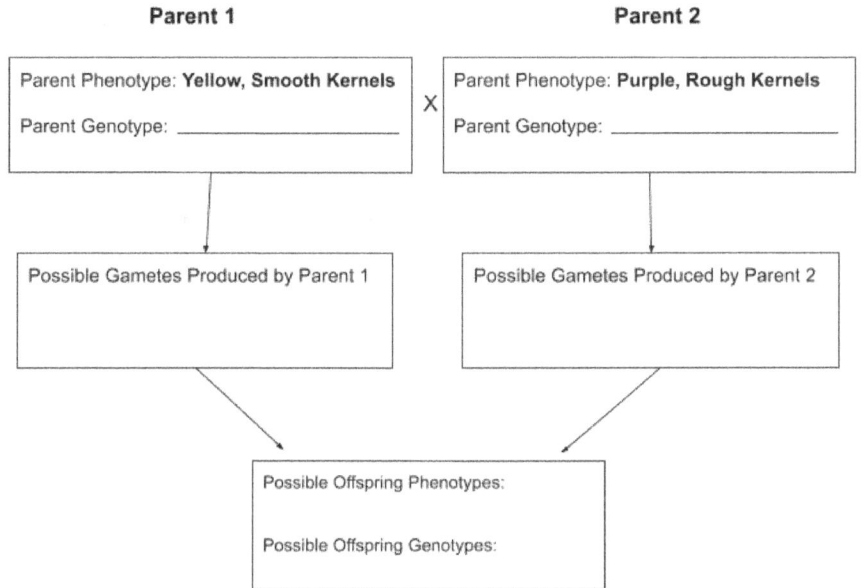

4. **Complete figure 7.4** by adding the genotype of each F_1 parent and the gametes each of those plants produces.
    a. The kernels on the cobs observed in this exercise are representative of the **F_2 generation**.
    b. Observe the cob to determine the four possible phenotype combinations (including both color and structure) produced by this cross.
    c. Possible genotypes will be determined in the next step.

Figure 7.4. Diagram of the Cross Creating the F2 Generation (two traits)

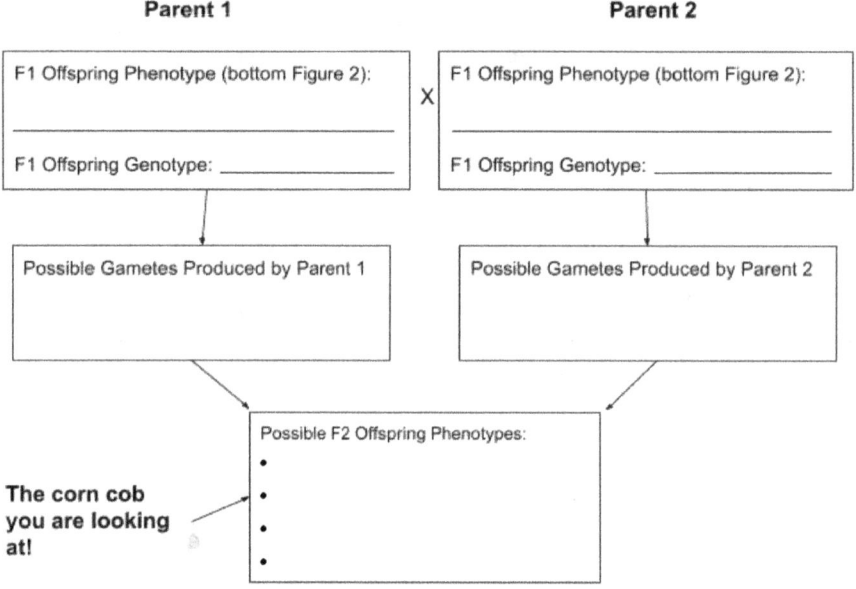

5. Determine the possible genotypes of the F_2 offspring by **completing the Punnett Square (figure 7.5)**.
    a. Add the gametes listed for Parent 1 in figure 7.4 to the left side of the Punnett Square (each gamete should correspond to one row in the table.
    b. Add the gametes listed for Parent 2 in figure 7.4 to the top of the Punnett Square. Each gamete should correspond to one column in the table.
    c. Complete each cell in the Punnett Square by "multiplying" the gamete for each row by the gamete for each column.
    d. Place the two copies of each letter together. If a trait is heterozygous, place the capital letter in front of the lowercase letter.
6. Use table 7.5 to "decode" the genotype and **write the phenotype below the genotype** in each cell of the Punnett Square.

Figure 7.5. Punnett Square Displaying the Possible Genotypes and Phenotypes of the F₂ Offspring (2 traits)

7. List the four possible phenotypes produced by this cross (from figure 7.4 or 7.5) in table 7.5.
    a. Determine the percentage of the offspring that should exhibit each phenotype.
    b. Divide the number of squares that exhibit each phenotype by 16 (the number of squares in the table).
    c. Multiply by 100 (to get the percentage).

Table 7.5. Results of Cross Following Two Traits Based on the Punnett Square

Possible Phenotypes	Number with this phenotype in Punnett Square	Percent of Offspring Predicted to have this Phenotype
Totals:	16	100

Genetics 121

8. Use your calculations in table 7.5 to create a hypothesis and prediction for the results of this experiment (cross).

Q10. What is the hypothesis? (Return to Exercise 7.1.A if you need a hint about how to format your hypothesis. This is a different hypothesis focusing on a different one of Mendel's laws)

Q11. What do you predict the outcome of the cross will be (write it out like we did for Exercise 7.1.A)?

9. Determine and record the number of individuals with each phenotype combination on the provided corn cob in table 7.6. **Do not damage the corn cobs in any way (e.g., removing kernels, marking them with a pen or sharp object).**
    a. An "X" has been marked at the end of one row of kernels at the larger end of the corncob. Use this as your starting row.
    b. Cut a length of blue tape equivalent to the length of your corncob.
    c. Align the tape with the bottom of the row marked with the X.
    d. Use the provided laboratory counters to record the phenotype of each kernel in that row. This usually works best if one teammate reads the phenotypes and a second teammate clicks the counter.
    e. When you reach the end of the row, move the tape so it covers that row and lines up below the next row. Count the next row.
    f. Continue counting until you get back to the X marking the first row.
10. Gather additional data from the other groups in your class.
11. Determine the percent of offspring represented by each phenotype combination (record in table 7.6).

Table 7.6. Results of the Dihybrid Cross

Phenotype combination	Your Group	Group 2	Group 3	Group 4	Group 5	Group 6	Totals	Percent
						Total # kernels:		100%

Q12. How does your data compare to your prediction? Is your hypothesis correct? Explain.

Q13. Why do we look at the percentage of offspring of each phenotype rather than genotype?

Q14. In this lab, you were asked to gather data from several other groups before calculating your results. Why would it be important to have a large pool of data for a genetics experiment?

Exercise 7.2. Following Human Genetics Using Pedigrees

Many human diseases are caused by mutations. Studying the inheritance of a particular disease-causing mutation is challenging due to the lifespan and low reproductive rates of humans.

Pedigrees depict the inheritance of a trait in a set of related individuals. Pedigrees can be used to determine the probability that a particular individual carries one or more alleles of a gene.

Figure 7.6. Understanding Human Pedigrees

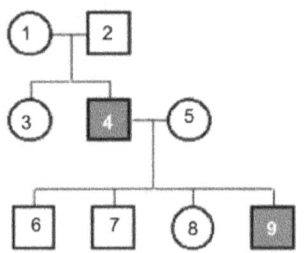

Understanding a pedigree:
- Males are represented by **squares**
- Females are represented by **circles**
- The shape is **shaded in** if that individual exhibits the trait being followed
- **Horizontal lines** between a male and female represent mating
- **Vertical lines** extending down from a couple represent that couple's children
- The youngest individuals are always at the bottom of the pedigree while the oldest are always at the top.

To interpret a pedigree, start with what you know.

- ❑ In figure 7.6, individual 9 exhibits the trait as does his father, individual 4. Individual 9's mother and siblings do not exhibit the trait. *This tells us that the trait is inherited as a recessive allele.*
- ❑ If the gene is represented by the letter A, individuals 4 and 9 must be homozygous recessive or "aa."
- ❑ Since individual 4 can only produce "a" gametes, all of his kids must have at least one "a." That means individual 6 (does not exhibit trait) has the genotype "Aa."

Q15. Take a minute to assign genotypes to all the individuals in figure 7.6. If more than one genotype is possible, list all the possibilities!

Q16. How many children were produced by the oldest members of this family? _____

Q17. How many of those children were girls? _____

Q18. How many of those children exhibited the trait being followed? _____

Q19. Is the trait being studied a dominant or recessive allele? _____

Q20. What are the genotypes of the oldest members of this family? _____

Hemoglobin is a protein found in red blood cells (RBCs) that transports oxygen throughout the body. Sickle cell disease (also called sickle cell anemia) is caused by a genetic mutation in the DNA sequence that codes for part of the hemoglobin protein. The mutation causes a change in amino acid sequence that causes the protein to clump together within the RBC. This clumping causes the RBC to assume an abnormal "sickle" shape.

> **Tie in!** We made models of the oxygen binding region of the HBA and HBS alleles during the gene expression lab. Recall how the change in a single amino acid caused the entire protein to change shape. Revisit your models (or your photo of them) to help form a link between genotype and phenotype in this example.

- ❏ Individuals who are homozygous for the normal hemoglobin allele (HH) produce normal hemoglobin and do not exhibit sickle-shaped RBCs.
- ❏ Individuals who are heterozygous (Hh) are said to have **sickle cell trait**. These individuals produce some of the mutated protein and do not normally exhibit symptoms. Heterozygous individuals do exhibit some symptoms of sickle cell anemia in low-oxygen environments.
- ❏ Individuals that are homozygous for the sickle cell allele (hh) produce only sickle-shaped RBCs, which tend to block blood flow, causing pain, serious infections, and organ damage.

Procedure

1. Answer the following questions regarding genetics and pedigrees. Attach additional paper as necessary to show your work.

Q21. Two people with sickle cell trait have children.

What are the genotypes of the parents? _____

What is the genetic makeup of the gametes the mother can produce? _____

What is the genetic makeup of the gametes the father can produce? _____

What is the chance that a child will have sickle cell disease?

What is the chance that a child will be resistant to mosquitos?

Q22. An individual who has sickle cell trait has children with an individual who does not have any copies of the sickle cell allele.

What are the genotypes of the parents?

What are the genotype and phenotype ratios of the possible offspring?

What are the chances that this couple will produce a child with sickle cell disease?

If this couple lives in the lowlands of East Africa, what are the chances that one of their children would be resistant to malaria if exposed to the malaria parasite?

Q23. The following pedigree traces sickle cell disease through four generations of a family living in New York City. Use the pedigree in figure 7.7 to answer the following questions.

Figure 7.7. Human Pedigree for Q23

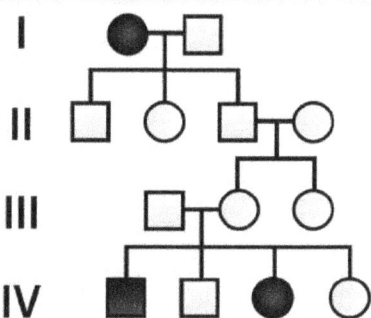

What is the genotype of the mother in the first generation?

What are the possible genotypes of the father in the first generation?

What can you say about the genotype of all the children of the couple in the first generation? Explain your answer.

Based on where the family resides, would the genotype of the individuals in the second generation be considered a disadvantage? Why or why not?

What are the genotypes of the parents in the third generation? Explain how you know.

Q24. The following pedigree traces sickle cell disease through four generations of a family living in the highlands of eastern Africa. Use the pedigree in figure 7.8 to answer the following questions.

Figure 7.8. Human Pedigree for Q24

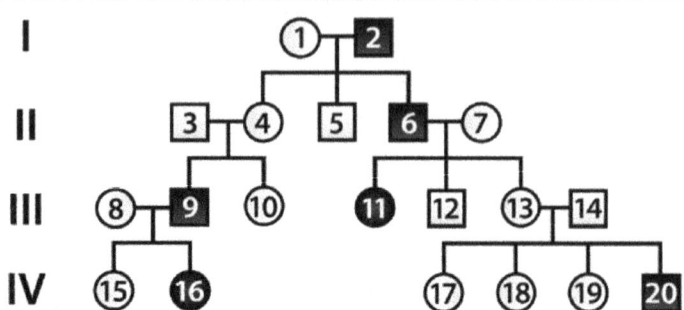

What are the genotypes of the following individuals? (If more than one genotype pertains, include all possibilities.)

- ❑ Individual 1: _____
- ❑ Individual 2: _____
- ❑ Individual 7: _____
- ❑ Individual 10: _____
- ❑ Individual 13: _____
- ❑ Individual 17: _____

If individuals 13 and 14 have another child, what are the chances that the child will have sickle cell disease?

Based on where this family lives, is the sickle cell trait genotype a genetic advantage? Explain.

If individuals 8 and 9 have four more children, what are the chances that two of the children will be homozygous for normal RBCs? Explain why.

128 Cell Biology, Genetics, and Evolution

Name: _____ Lab Time: _____ Due: _____

7 | Applying What You've Learned in the Genetics Lab

Solve the following genetics problems. Show your work. Circle your final answer.

1. An organism has the genotype Aa.

    ❑ How many genes are being studied in this organism?

    ❑ Is this organism haploid or diploid?

    ❑ What are the possible gametes that this organism can make?

2. An organism has the genotype AaBB.

    ❑ How many genes are being studied in this organism?

    ❑ Is this organism haploid or diploid?

❑ What are the possible gametes that this organism can make?

3. What is the best way to determine which of two different alleles for a single gene is dominant?

4. When creating a Punnett Square, what goes on the top, along the side, and into the cells of the square?

5. Can two unaffected individuals have a child who has a genetic disorder (genetically related to both parents)? Explain your answer.

Name: _____   Lab Time: _____   Due: _____

7 | Genetics Problems

Attach additional paper as necessary to show your work.

1. In peas, the yellow seed color is dominant to the green seed color. What colors and proportions of offspring would result from a cross between a heterozygous yellow-seeded plant and a homozygous green-seeded plant?

2. Cystic fibrosis is a common recessive disorder. What is the probability that two normal parents, both of whom are heterozygous for cystic fibrosis, will have an affected child?

3. Huntington's disease results from the presence of a dominant allele, and the disease does not show up until middle age (40s). Could two parents who do not show signs of Huntington's in their 20s have children with Huntington's? Why or why not?

4. In tomatoes, red fruit is dominant over yellow fruit, while tall vines are dominant over short vines. A tomato breeder has pure (homozygous) varieties of yellow-fruited, tall plants and red-fruited, short plants:

   ❏ If the breeder crosses the two varieties that he has, what will be the appearance of the F1?

   ❏ If a breeder crosses a tomato that is heterozygous for both height and fruit color with a tomato that is homozygous recessive for height and heterozygous for color, what percent of the offspring should have short vines and yellow fruit?

5. The following pedigree traces sickle cell disease through three generations of a family (generations labeled with Roman numerals on the left). Use the pedigree in figure 7.9 to answer the following questions.

Figure 7.9. Human Pedigree for Sickle Cell Disease

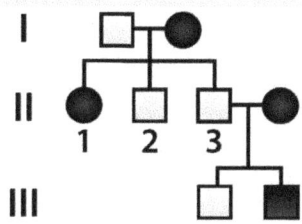

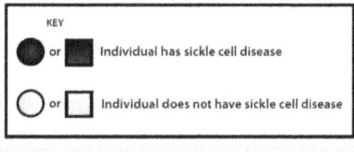

What is the genotype of the father in the first generation?

What is the genotype of the daughter in the second generation?

What is the genotype of individual 3 in the second generation? How do you know?

If the couple in the second generation has another child, what are the chances the child will have the following?

- ❑ Sickle cell disease: _____
- ❑ Sickle cell trait: _____
- ❑ Normal hemoglobin: _____

If the entire family moves to the lowlands of East Africa, four of the five males in the pedigree will have two genetic advantages over the other individuals in the family. Explain the two advantages.

Name: _____ Lab Time: _____ Due: _____

8 | Natural Selection Pre-lab

Exercise 8.1. Darwin's Finches

When Charles Darwin explored the Galapagos Islands, he made observations about the features of organisms that lived on different islands. Those differences seemed to be related to the environment.

Figure 8.1. Darwin's Drawings of Galapagos Finches

1. Geospiza magnirostris.
2. Geospiza fortis.
3. Geospiza parvula.
4. Certhidea olivasea.

1. Use figure 8.1 to identify and list three different examples of phenotype variation in Galapagos finches.

2. Choose one of those examples and explain how variation in that phenotype might contribute to finch survival.

Exercise 8.2. DNA Fingerprinting Review

Return to Lab 3: DNA Fingerprinting (p. 41) or watch the video to review restriction enzyme digest and agarose gel electrophoresis. Scan the QR code or visit https://go.chemeketa.edu/gelelectrophoresis to access the video.

3. The restriction enzyme EcoR1 has the restriction site GAATTAC. Underline or highlight the EcoR1 restriction sites in DNA Sequence 1 and DNA Sequence 2 in figure 8.2

Figure 8.2. Sample DNA Sequences. Both sequences are 35 base pairs (bp) long. *Made with Biorender.*

DNA Sequence 1   AGTACCGAATTCACTTTAAGAATTCCGATTAGATT

DNA Sequence 2   AGTACCGTATTCACTTTAAGACTTCCGATTAGATT

4. How many pieces of DNA are produced by restriction enzyme digestion of each sequence with EcoR1?

    a. DNA Sequence 1:
    b. DNA Sequence 2:

5. What are the sizes in base pairs of the DNA fragments produced by the EcoR1 digest?

    a. DNA Sequence 1:
    b. DNA Sequence 2:

6. What is the charge of a DNA molecule?

    ☐ Positive
    ☐ Negative

7. What force moves the DNA through the gel in agarose gel electrophoresis?

8. By what properties are DNA fragments separated in agarose gel electrophoresis?

9. What size of DNA fragment moves faster in agarose gel electrophoresis?

    ☐ Large
    ☐ Small

10. Indicate the orientation of the agarose gel in figure 8.3 by adding a ✚ or ▬ sign to the circles indicated by the electrical field arrows.

    Figure 8.3. Results of Agarose Gel Electrophoresis of Digested DNA Sequences in figure 8.2. *Made with Biorender.*

    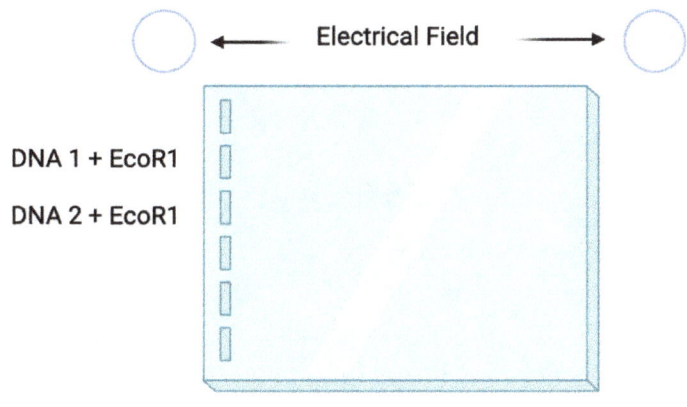

11. Complete the gel in figure 8.3 by adding the bands representing the fragments you calculated for DNA 1 and DNA 2 to the gel.

Name: _____ Lab Time: _____ Due: _____

8 Natural Selection

By the end of this lab, you should be able to:

- ❑ Explain the process of natural selection.
- ❑ Describe the impact of the environment on adaptation.
- ❑ Calculate the frequencies of phenotypes and genotypes within a population.
- ❑ Graph population data.

Exercise 8.3. Natural Selection in Galapagos Finches*

Charles Darwin's understanding of evolution by natural selection developed over many years and many experiences. The most well-known example of his journeys, perhaps, is his trip to the Galapagos Islands on the HMS Beagle. This was where Darwin encountered a variety of different examples of natural selection. Finches were just one of the species Darwin observed, and they played a minor role in the development of his theory, but they are the example we most associate with him. Figure 8.4 includes images of two of those species.

Figure 8.4. Two different medium ground finches (A, B). Seeds that finches eat from the genuses Euphorbia (C), Sida (D), and Tribulus (E). Plants and rocks on an island in the Galápagos where finches live (F).

* Based on *Beaks as Tools: Selective Advantage in Changing Environments* (2021) www.biointeractive.org

In this exercise, we will do what Darwin did—make observations and test models to get a better understanding of how the process of natural selection works.

A. Using Finch Observations to Make Predictions

The scientific process is based on observation, asking questions and forming hypotheses. In this exercise, you will make observations of images of Galapagos Finches, the food they consume and the environment they live in. We will use these observations to predict how the birds interact with the different foods available to them.

Procedure

1. Complete table 8.1 by adding at least three observations to each column about the finches, the seeds, and the environment in figure 8.4.
2. Complete table 8.2 by using your observations to predict how each type of finch in figures 8.4A and 8.4B will interact with each type of seed (figures 8.4C, 8.4D, and 8.4E).

Table 8.1. Observations of Galapagos Finches, the Foods They Eat, and the Environment They Live In

Finches (A, B)	Seeds (C, D, E)	Environment (F)

Table 8.2. Predictions of How Galapagos Finches Will Interact with the Different Types of Seeds in Their Environment

Finch Type	Predicted Interactions Between Finches and the Three Seed Types Found in Their Habitats		
	Seed Type C	Seed Type D	Seed Type E
A			
B			

B. Using Finch Observations to Make a Model

The environment is constantly changing, and according to natural selection, individuals who are better able to survive an environmental change are more likely to produce offspring. Those offspring carry the genes that helped their parents survive environmental change. Because those genes are helpful for survival, they will become more common over time (as long as the same environmental pressures are driving evolution). In this model, we will use a box, substrate, rice, beans, tweezers, and pliers to represent finches, seeds, and their environment.

Q1. Draw lines to match the parts of our model on the left to the role they play in our model on the right.

Model Component	Role
Beans	Environment
Box with substrate	Finch Species
Pliers	Seeds
Rice	
Tweezers	

Procedure

1. Try out the various components of the model.
2. Practice picking up the seeds with the tweezers and the pliers. Keep the following in mind:
   a. To eat a **small seed** (rice), just pick it up and swallow it (put it in a cup).
   b. To eat a **large seed** (bean), you must crush it before swallowing it. Large seeds normally have thick coats that must be broken to obtain the nutrition inside.
3. Put seeds into the environment you will be using and try to collect seeds with both tools.

Q2. Compare the efficiency of each tool in gathering each type of seed.

Q3. Compare the efficiency of each tool in crushing the big seeds (beans).

Q4. How does the environment affect the efficiency of gathering each type of seed?

Q5. All models have limitations. Describe one aspect of finch habitat that cannot be accounted for in this model.

Q6. Complete table 8.3 by using what you have observed about the tools, seeds, and environment to predict what will happen in the model under three different environmental conditions.

Table 8.3.

Environmental Condition	Small beak (tweezers)	Large beak (pliers)
Both small and large seeds are available.		
Only large seeds are available.		
Only small seeds are available.		

C. Galapagos Islands Before 1977

Both small and large seeds are available.

Procedure

1. Designate two members of your group to be "finches." Finches will use a tool (tweezers or pliers) to "eat" as many seeds as possible under a time limit.
2. The remaining members of your group will be "observers." Observers will:
   a. add seeds to the environment
   b. oversee and time the feeding trials and count "eaten" seeds
   c. make sure that large seeds, if any, are crushed and counted correctly
3. Each finch should take a different tool (either tweezers or pliers)
4. The observers should sprinkle a **large scoop of large seeds** (beans; large beaker) and a **small scoop of small seeds** (rice; small beaker) into your environment (box with substrate).
5. Gather seeds for 30 seconds.
   a. An observer should start the timer.
   b. The finches should pick up as many seeds as they can during the 30 seconds and drop them into their cups. Remember that large seeds must be crushed before putting them in the cup.
   c. An observer should count the number of large seeds gathered and crushed before they are put into the cup. It can be hard to count crushed seeds after the fact!
   d. At the end of the 30-second gathering period, count the number of small and large seeds gathered. Record your data in table 8.4.
6. Remove the seeds from your cup. Do not put the seeds back into the environment. Your trials encompass an entire season of feeding, so once a seed is consumed, it is no longer available to eat.
7. Repeat seed gathering 4 more times, recording your data into the Before 1977 (purple) columns of table 8.4.
8. At the end of the season, add up the total number of seeds consumed by each type of finch (total row).
9. Remove all seeds from the environment.

Table 8.4. Number of Seeds Consumed by Small-Beaked and Large-Beaked Finches in Different Environments.
All data should include the total number of seeds of any size collected.

Results	Before 1977 (small and large seeds)		1977–1983 (only large seeds)		After 1983 (only small seeds)	
	"Beak" type		"Beak" type		"Beak" type	
	Small	Large	Small	Large	Small	Large
Trial 1						
Trial 2						
Trial 3						
Trial 4						
Trial 5						
Total						
Mean						
80% limit						

Q7. What do your data tell you about the medium ground finch feeding on the Galapagos Islands before 1977?

Q8. Do bigger seeds have more nutritional value? If so, a finch that eats small seeds would need to eat far more seeds than a finch with a large beak eats. Does your data support your answer?

D. Galapagos Islands 1977–1983
Only large seeds are available.

Procedure
1. Designate two members of your group as finches and the other two as observers.
2. Each finch should take a different tool (either tweezers or pliers)
3. The observers should sprinkle ½ **of a measuring cup of beans (large seeds)** into the environment.
4. Gather seeds for 30 seconds using the same method described in the last model. Remember that large seeds have to be crushed before they are consumed!
5. Remove the seeds from your cup. Do not put the seeds back into the environment.
6. Repeat seed gathering 4 more times, recording your data into the 1977–1983 (blue) columns of table 8.4.
7. At the end of the season, add up the total number of seeds consumed by each type of finch (total row).
8. Remove all seeds from the environment.

Q9. How did finch feeding change between 1972 and 1983?

E. Galapagos Islands After 1983
Only small seeds are available.

Procedure
1. Designate two members of your group as finches and the other two as observers.
2. Each finch should take a different tool (either tweezers or pliers)
3. The observers should sprinkle **½ of a measuring spoon of rice (small seeds)** into the environment.
4. Gather seeds for 30 seconds using the same method described in the last model.
5. Remove the seeds from your cup. Do not put the seeds back into the environment.
6. Repeat seed gathering 4 more times, recording your data into the After 1983 (orange) columns of table 8.4.
7. At the end of the season, add up the total number of seeds consumed by each type of finch (total row).
8. Remove all seeds from the environment.

Q10. How did finch feeding change after 1983?

F. Analyzing Finch Survival
In order to survive, a finch must gather a certain amount of food. For the sake of this exercise, assume the following:

- In the first simulation we performed (1977), all finches got enough food to be full. This is its **free-feeding amount**.
- A finch must consume 80% of its free-feeding amount to survive. This is the **80% limit**.

Procedure
1. Calculate the mean (average) number of seeds collected by each beak type under each condition. Record the results in the **"Mean"** row of table 8.4.
2. Calculate the 80% limit for finches with small beaks (0.8 × mean number of seeds collected). **Record your answer in the "80% limit" row for the small beak** in the 1977 column of table 8.4.

3. Calculate the 80% limit for finches with large beaks. Record your answer in the **"80% limit" row for the large beak** in the 1977 column of table 8.4.
4. Compare the 80% limits you calculated with the mean number of seeds that your finches collected in the 1977–1983 and the After 1983 trial.
    a. If the calculated mean for a column is larger than or equal to the 80% limit, the finch survived.
    b. If the calculated mean for a column is smaller than the 80% limit, the finch dies.
    c. **Mark the finches that died with a cross in the "mean" row.**

Q11. In which simulation (all large seeds and/or all small seeds) did finches die?

Q12. Under what conditions is it beneficial to be a large-beaked finch on the Galapagos Islands?

Q13. Under what conditions is it beneficial to be a small-beaked finch on the Galapagos Islands?

Exercise 8.4. Is Beak Size Inherited?**

Although Darwin's study of the organisms inhabiting the Galapagos Islands took place more than 150 years ago, ground finches remain a model system for studying evolution. The genomes of all Galapagos finch species were published by 2016, allowing researchers to investigate the role of DNA changes in finch evolution. Comparing finch DNA to that of other organisms allowed researchers to identify two finch genes that were similar to genes known to be involved in the formation of facial features in mammals. Later, scientists identified one of those genes, *HMGA2*, as having small differences in DNA sequence in ground finches exhibiting different beak sizes. Could this gene regulate beak size?

Q14. How did the finch scientists use what was already known about mammals to locate genes that might be involved in beak development?

Q15. If both finches and mammals use the same genes for the development of facial features, is it more likely that the genes evolved separately in finches and mammals or that the genes evolved before the evolutionary paths of finches and mammals separated?

Q16. Why is it important to know if a trait like beak size is determined by a gene?

** Adapted from "Genetic Flight Plan: Navigating Galapagos Finch Diversity," Edvotek® Kit 921

By comparing the DNA sequence of the *HMGA2* gene of 60 ground finches, researchers identified a region of the potential beak size gene that varied based on phenotype. Birds that were homozygous for one version of that *HMGA2* had large beaks, birds that were homozygous for the other version of *HMGA2* had small beaks and those that were heterozygous had medium beaks.

Q17. What type of dominance is exhibited by the gene that may be associated with regulating beak size? Explain your answer.

In Lab 3: DNA Fingerprinting, we learned that **restriction enzymes** cut the DNA of different individuals in different locations due to differences in their DNA sequences. Each restriction enzyme cuts the DNA at a specific and predictable location (**restriction site**) in the genome. Molecular biologists use restriction enzymes as "scissors" to cut DNA for a variety of purposes including differentiating between the different alleles of a gene.

The difference between the two alleles of the *HMGA2* includes a difference in the presence of the restriction site for EcoR1.

- ❏ The suspected large allele lacks a restriction site for EcoR1
- ❏ The suspected small allele contains a restriction site for EcoR1.

To determine if *HMGA2* controls beak size, scientists used specialized techniques to make millions of copies of the *HMGA2* gene in finches with big beaks and then in finches with small beaks. Then, they incubated each DNA sample with the restriction enzyme EcoR1.

If the suspected gene controls beak size, then the DNA different restriction patterns should correlate with different finch beak sizes.

A. Practice Loading a Gel

Take a few minutes to review using a micropipette and practice loading a gel. Everyone in your group should take a turn at practicing.

Procedure
1. Fill the dish containing your practice gel with tap water until the gel is fully covered by the water. The water should be deep enough that its surface is mirror-smooth.
2. Set a micropipette to 10 microliters and add a tip. Everyone in your group will use the same tip.
3. Use the micropipette to remove 10 microliters of practice dye from the "PD" tube.
4. Slowly drop the dye into the well.

After everyone has practiced:

- ❑ Return the PD tube to your instructor (assuming there is remaining practice dye).
- ❑ Rinse off the silicone practice gel and its container and return them to your student kit.

B. Load and Run the Gel

Procedure
1. Place your agarose gel in the electrophoresis apparatus. Check your pre-lab if you need to review orientation.
2. Place the electrophoresis apparatus near the power source that you will be sharing with the other team at your lab bench. Be sure that the cables on your apparatus reach the power source, as you cannot move the gel once the samples have been loaded.
3. Add enough 1X TAE to cover the surface of your agarose gel. The solution should fill the wells on both sides of the gel and cover the gel. The surface of the solution should be smooth. Check with your instructor before continuing.
4. Tap the sample strip firmly on the counter to concentrate the contents at the bottom of the tube. **DO NOT remove the foil strip from the tubes.**
5. Set your micropipette to 35 microliters.
6. Using a different tip for each sample, load your DNA samples into the gel.
    a. Puncture the foil seal of the tube with the pipette tip.
    b. Push down to the soft stop on the micropipette.
    c. Insert the tip into the tube and draw up the sample
    d. Load your samples in order in the remaining lanes (see table 8.5)

Table 8.5 Recommended Order of Samples in Gel

Lane	Tube	Sample Contents
1	A	DNA marker
2	B	Large beak DNA control
3	C	Samll beak DNA control
4	None	Empty
5	D	DNA from Finch 1, large beak
6	E	DNA from Finch 2, medium beak
7	F	DNA from Finch 3, small beak

Carefully slide the cover onto the electrophoresis apparatus and plug your gel into the power source.

7. When the other team using your power source has attached their gel, turn on the power to high. If everything is set up correctly, you should see tiny bubbles streaming up the short sides of the apparatus.
8. Allow the gel to run for approximately 20-30 min. The gel is done when the visible dye line is about 3.5 cm from the wells.

Q18. What is the role of the large and small beak control samples (samples B and C) in this experiment? When was the DNA in these samples likely collected?

Remember!
- The same restriction enzyme is used on all DNA samples in this experiment.
- The restriction enzyme does not cut the large beak allele.
- The restriction enzyme cuts the small beak allele twice.

Q19. Use the information from table 8.5 to complete figure 8.5 to predict the results of your gel. *Hint: Remember that each finch contains two alleles for each gene.*

Figure 8.5 Predicted Results of DNA Gel Electrophoresis. The large allele lacks a restriction site and the small allele has one restriction site. *Made with Biorender*

B   Large Beak Control DNA
C   Small Beak Control DNA
D   Finch 1 with a large beak
E   Finch 2 with a medium beak
F   Finch 3 with a small beak

Q20. Explain how your results will allow you to determine whether the suspected gene controls beak size?

C. View the Gel

Procedure

1. After the gel has finished running, remove it from the electrophoresis apparatus and carefully carry it to one of the viewing stations around the room.
2. Open the lid of the Blue View Transilluminator and place your gel on the blue surface. Close the lid so that the transparent orange plastic is between you and your gel.
3. Turn on the transilluminator by turning the knob on the right-hand side. Your bands will become visible.
4. Complete figure 8.6 by adding the bands you see in each lane. You may also want to take a photograph of the gel.

Figure 8.6. Results of DNA Gel Electrophoresis of Finch DNA Samples

Q21. Do your results match the predictions you made in figure 8.5? Why or why not?

Q22. Based on your data, does this gene appear to control beak size?

Natural Selection 153

8

Name: _____ Lab Time: _____ Due: _____

8 Applying What You've Learned in the Natural Selection Lab

1. Do your data suggest that individual finches change in response to the environment or that the population changes in response to changes in the environment? Explain your answer.

2. Do finches evolve different-sized beaks as needed for survival? Explain.

3. A tropical storm moves over the Galapagos Islands at peak seed ripeness. The high wind blows many small seeds out to sea, and large seeds drop to the ground. What do you predict will happen to the beak size of medium-ground finches? Why?

4. Due to a summer of sufficient rainfall, all the plants on the Galapagos Islands are producing a bumper crop. Big and small seeds are plentiful. What do you predict will happen to the beak size of medium-ground finches? Why?

5. If you were to analyze the beak-size gene from 20 ground finches during a drought would you expect find:

    ☐ All the finches to have DNA like Finch 1 (Lane 4 of the gel)
    ☐ All the finches to have DNA like Finch 2 (Lane 5 of the gel)
    ☐ All the finches to have DNA like Finch 3 (Lane 6 of the gel)
    ☐ Some individuals with DNA like Finch 2 and some with DNA like Finch 3
    ☐ An equal number of samples like Finches 1, 2, and 3.

    Explain your answer.

6. The gel shown in figure 8.7 is the result of restriction enzyme digestion and electrophoresis of the beak size gene from 12 individual ground finches.

    Figure 8.7. Agarose Gel Electrophoresis of Restriction Enzyme Digest of *HMGA2* Gene. *Made with Biorender.*

Which of the test groups from Exercise 8.3 is best represented by the gel shown in figure 8.7?

- ❏ The Galapagos Island ground finch population before 1977
- ❏ The Galapagos Island ground finch population from 1977–1983
- ❏ The Galapagos Island ground finch population after 1983?

Explain your answer.

Acknowledgments

Unless otherwise stated, figures created and copyrighted by Jennifer Schramm or Chemeketa Press © 2025. All rights reserved.

Figure 5.2. "The Genetic Code," by Genome.gov. Adapted by Chemeketa Press.

Figure 6.4. "Cross Section of a Lily Anther," by Bruce Kirchoff. Flickr. Wikimedia Commons.

Figure 7.7. "Human Pedigree for Q23," adapted by Jennifer Schramm from HHMI BioInteractive, "Mendelian Genetics, Probability, Pedigrees and Chi-Square Statistics."

Figure 7.8. "Human Pedigree for Q24," adapted by Jennifer Schramm from HHMI BioInteractive, "Sickle Cell Disease Storyline" (2024).

Figure 7.9. "Human Pedigree for Sickle Cell Disease," adapted by Jennifer Schramm from HHMI BioInteractive, "Mendelian Genetics, Probability, Pedigrees and Chi-Square Statistics."

Figure 8.1. "Darwin's Drawings of Galapagos Finches" by Charles Darwin, London: J Murray, 1845. Biodiversity Heritage Library. Contributed by MBLWHOI Library.

Figure 8.4. "Beaks as Tools: Selective Advantage in Changing Environments" (2021). HHMI Biointeractive.

www.ingramcontent.com/pod-product-compliance
Lightning Source LLC
Chambersburg PA
CBHW081713070825
30679CB00001BA/3